T0222598

Advanced Structural Materials

MATERIALS RESEARCH SOCIETY
SYMPOSIUM PROCEEDINGS VOLUME 1243

Advanced Structural Materials

Symposium held August 16–21, 2009, Cancun, Mexico

EDITORS:

Hector A. Calderon

ESFM-Instituto Politécnico Nacional
UPALM-Zacatenco, Mexico D.F., Mexico

Armando Salinas-Rodríguez

CINVESTAV-IPN, Unidad Saltillo
Ramos Arizpe, Coahuila, Mexico

Heberto Balmori-Ramirez

Instituto Politécnico Nacional
UPALM-Zacatenco, Mexico D.F., Mexico

J. Gerardo Cabañas-Moreno

CNNMT-Instituto Politécnico Nacional
UPALM-Zacatenco, Mexico D.F., Mexico

Kozo Ishizaki

Nagaoka University of Technology
Nagaoka, Japan

Materials Research Society
Warrendale, Pennsylvania

CAMBRIDGE UNIVERSITY PRESS
Cambridge, New York, Melbourne, Madrid, Cape Town,
Singapore, São Paulo, Delhi, Mexico City

Cambridge University Press
32 Avenue of the Americas, New York NY 10013-2473, USA

Published in the United States of America by Cambridge University Press, New York

www.cambridge.org
Information on this title: www.cambridge.org/9781107406797

Materials Research Society
506 Keystone Drive, Warrendale, PA 15086
http://www.mrs.org

© Materials Research Society 2010

This publication is in copyright. Subject to statutory exception
and to the provisions of relevant collective licensing agreements,
no reproduction of any part may take place without the written
permission of Cambridge University Press.

This publication has been registered with Copyright Clearance Center, Inc.
For further information please contact the Copyright Clearance Center,
Salem, Massachusetts.

First published 2010
First paperback edition 2012

Single article reprints from this publication are available through
University Microfilms Inc., 300 North Zeeb Road, Ann Arbor, MI 48106

CODEN: MRSPDH

ISBN 978-1-605-11220-6 Hardback
ISBN 978-1-107-40679-7 Paperback

Cambridge University Press has no responsibility for the persistence or
accuracy of URLs for external or third-party internet websites referred to in
this publication, and does not guarantee that any content on such websites is,
or will remain, accurate or appropriate.

CONTENTS

PREFACE

This volume is a compilation of papers that were presented at Symposium 5, "Advanced Structural Materials," that was held at the XVIII International Materials Research Congress 2009 that took place August 16–21, 2009 in Cancun, Mexico. This symposium was organized in collaboration with the Materials Research Society (MRS). The symposium was devoted to fundamental and technological applications of structural materials, and continued the tradition of providing a forum for scientists from various backgrounds with a common interest in the development and use of structural materials to come together and share their findings and expertise.

The papers contained in this volume are a collection of invited and contributed papers. This year, the symposium was attended by participants from France, India, Japan, Mexico, Portugal, the United States and Spain. All papers have been thoroughly reviewed by at least two referees and edited to the standards of MRS. We are grateful to all those referees who, by their comments and constructive criticism, helped to improve the finally printed papers, and to all the authors who made additional efforts to prepare their manuscripts.

The "Advanced Structural Materials" symposium has been held for the last 12 years with the objective of presenting overview and recent investigations related to advanced structural metallic, ceramic and composite materials. The topics include innovative processing, phase transformations, mechanical properties, oxidation resistance, modeling and the relationship between processing, microstructure and mechanical behavior. Additionally, papers on industrial application and metrology are included.

The organizing committee gratefully acknowledges the enthusiastic cooperation of all of the symposium participants, as well as the kind acceptance of the editorial committee of MRS to publish these proceedings. The organizing committee and the Editors wish to thank the Instituto Politécnico Nacional, Mexico (COFAA, Secretaria de Investigación y Posgrado) and the Centro de Investigación y de Estudios Avanzados del Instituto Politécnico Nacional for their financial support. We hope that all readers will come to consider the "Advanced Structural Materials" symposium at Cancun a suitable forum to present results of their recent research and experience.

Heberto Balmori-Ramírez
J. Gerardo Cabañas-Moreno
Hector A. Calderon-Benavides
Kozo Ishizaki
Armando Salinas-Rodríguez

January 2010

MATERIALS RESEARCH SOCIETY SYMPOSIUM PROCEEDINGS

MATERIALS RESEARCH SOCIETY SYMPOSIUM PROCEEDINGS

Volume 1218E —Energy Harvesting—From Fundamentals to Devices, H. Radousky, J.D. Holbery, L.H. Lewis, F. Schmidt, 2010, ISBN 978-1-60511-191-9

Volume 1219E —Renewable Biomaterials and Bioenergy—Current Developments and Challenges, S. Erhan, S. Isobe, M. Misra, L. Liu, 2010, ISBN 978-1-60511-192-6

Volume 1220E —Green Chemistry in Research and Development of Advanced Materials, W.W. Yu, H. VanBenschoten, Y.A. Wang, 2010, ISBN 978-1-60511-193-3

Volume 1221E —Phonon Engineering for Enhanced Materials Solutions—Theory and Applications, 2010, ISBN 978-1-60511-194-0

Volume 1222 — Microelectromechanical Systems—Materials and Devices III, J. Bagdahn, N. Sheppard, K. Turner, S. Vengallatore, 2010, ISBN 978-1-60511-195-7

Volume 1223E —Metamaterials—From Modeling and Fabrication to Application, N. Engheta, J. L.-W. Li, R. Pachter, M. Tanielian, 2010, ISBN 978-1-60511-196-4

Volume 1224 — Mechanical Behavior at Small Scales—Experiments and Modeling, J. Lou, E. Lilleodden, B. Boyce, L. Lu, P.M. Derlet, D. Weygand, J. Li, M.D. Uchic, E. Le Bourhis, 2010, ISBN 978-1-60511-197-1

Volume 1225E —Multiscale Polycrystal Mechanics of Complex Microstructures, D. Raabe, R. Radovitzky, S.R. Kalidindi, M. Geers, 2010, ISBN 978-1-60511-198-8

Volume 1226E —Mechanochemistry in Materials Science, M. Scherge, S.L. Craig, N. Sottos, 2010, ISBN 978-1-60511-199-5

Volume 1227E —Multiscale Dynamics in Confining Systems, P. Levitz, R. Metzler, D. Reichman, 2010, ISBN 978-1-60511-200-8

Volume 1228E —Nanoscale Pattern Formation, E. Chason, R. Cuerno, J. Gray, K.-H. Heinig, 2010, ISBN 978-1-60511-201-5

Volume 1229E —Multiphysics Modeling in Materials Design, M. Asta, A. Umantsev, J. Neugebauer, 2010, ISBN 978-1-60511-202-2

Volume 1230E —Ultrafast Processes in Materials Science, A.M. Lindenberg, D. Reis, P. Fuoss, T. Tschentscher, B. Siwick, 2010, ISBN 978-1-60511-203-9

Volume 1231E —Advanced Microscopy and Spectroscopy Techniques for Imaging Materials with High Spatial Resolution, M. Rühle, L. Allard, J. Etheridge, D. Seidman, 2010, ISBN 978-1-60511-204-6

Volume 1232E —Dynamic Scanning Probes—Imaging, Characterization and Manipulation, R. Pérez, S. Jarvis, S. Morita, U.D. Schwarz, 2010, ISBN 978-1-60511-205-3

Volume 1233 — Materials Education, M.M. Patterson, E.D. Marshall, C.G. Wade, J.A. Nucci, D.J. Dunham, 2010, ISBN 978-1-60511-206-0

Volume 1234E —Responsive Gels and Biopolymer Assemblies, F. Horkay, N. Langrana, W. Richtering, 2010, ISBN 978-1-60511-207-7

Volume 1235E —Engineering Biomaterials for Regenerative Medicine, S. Bhatia, S. Bryant, J.A. Burdick, J.M. Karp, K. Walline, 2010, ISBN 978-1-60511-208-4

Volume 1236E —Biosurfaces and Biointerfaces, J.A. Garrido, E. Johnston, C. Werner, T. Boland, 2010, ISBN 978-1-60511-209-1

Volume 1237E —Nanobiotechnology and Nanobiophotonics—Opportunities and Challenges, 2010, ISBN 978-1-60511-210-7

Volume 1238E —Molecular Biomimetics and Materials Design, J. Harding, J. Evans, J. Elliott, R. Latour, 2010, ISBN 978-1-60511-211-4

Volume 1239 — Micro- and Nanoscale Processing of Biomaterials, R. Narayan, S. Jayasinghe, S. Jin, W. Mullins, D. Shi, 2010, ISBN 978-1-60511-212-1

Volume 1240E —Polymer Nanofibers—Fundamental Studies and Emerging Applications, 2010, ISBN 978-1-60511-213-8

Volume 1241E —Biological Imaging and Sensing using Nanoparticle Assemblies, A. Alexandrou, J. Cheon, H. Mattoussi, V. Rotello, 2010, ISBN 978-1-60511-214-5

Volume 1242 — Materials Characterization, R. Pérez Campos, A. Contreras Cuevas, R.A. Esparza Muñoz, 2010, ISBN 978-1-60511-219-0

Volume 1243 — Advanced Structural Materials, H.A. Calderon, A. Salinas-Rodríguez, H. Balmori-Ramirez, J.G. Cabañas-Moreno, K. Ishizaki, 2010, ISBN 978-1-60511-220-6

Prior Materials Research Society Symposium Proceedings available by contacting Materials Research Society

Mater. Res. Soc. Symp. Proc. Vol. 1243 © 2010 Materials Research Society

From Micro to Nanometric Grain Size CVD Diamond Tools

Flávia A. Almeida[1], Margarida Amaral[2], Ermelinda Salgueiredo[1], António J.S. Fernandes[2], Florinda M. Costa[2], Filipe J. Oliveira[1], Rui F. Silva[1]
[1]Ceramics Eng. Dept., CICECO, University of Aveiro, 3810-193 Aveiro, Portugal.
[2]Physics Dept., I3N, University of Aveiro, 3810-193 Aveiro, Portugal.

ABSTRACT

CVD diamond coated tools are developed for applications as different as turning of cemented carbides and bone drilling. The diamond films are deposited by Hot Filament Chemical Vapor Deposition (HFCVD), with grain sizes varying from conventional micrometric (12 μm) to nanometric (< 100 nm) and film thickness up to 50 μm. Silicon nitride (Si_3N_4) ceramics are chosen for the base material in order to guarantee maximal adhesion. Both the micrometric and nanometric CVD diamond grades endure the cemented carbide turning showing slight cratering, having flank wear as the main wear mode. However, nanocrystalline diamond present the best behavior regarding cutting forces (<150 N) and tool wear (KM=30 μm, KT=2 μm and VB=110 μm) and workpiece surface finishing (Ra=0.2 μm). In the case of the dental drilling experiments, a polymeric laminated test block is used to simulate the human mandible and maxilla. The temperature rise during drilling is monitored to prevent overheating above 42-47 °C that is known to cause tissue death and implant failure. It is possible to drill with a CVD diamond Si_3N_4 coated tool with significantly lower forces (fourfold smaller), lower rise in temperature (4°C less), lower spindle speeds (100 rpm) and higher infeed rates (30 mm/min), when compared to the commercial steel (AISI 420) drill bits.

INTRODUCTION

Microcrystalline CVD diamond cutting behaviour

The growth of diamond crystals by low pressure CVD technique was firstly documented in 1952, almost at the same time of the development of High Pressure High Temperature (HPHT) method of diamond manufacturing, by William Eversole of Union Carbide [1]. Nevertheless, the process was dismissed by most researches because the growth rate was very low, since graphite was co-deposited with the diamond leading to impure mixed phases [2]. In 1968, Angus and co-workers [3] were able to improve the diamond growth rate by including hydrogen in the carbon-containing gases. They discovered that the presence of atomic hydrogen during the deposition process lead to preferential etching of graphite, rather than of diamond.

The commercial availability of synthetic diamond cutting tools made by CVD route took place at the beginning of 1990s in two product forms: thick-film freestanding CVD diamond cutting tool tips and thin-film CVD diamond coatings [4,5]. The first form is produced by depositing a thick layer (from 150-1000 μm) of diamond on Si or Mo wafers, detached from the substrate and cleaned with acid solution. The next step is laser cutting of the free-standing diamond wafer in tool tips that are brazed onto a steel or cemented carbide body/insert tool. Finishing is made by grinding/polishing procedure of the tool cutting edge with the desired radius and edge angle.

With respect to thin-film coatings, these are made with fewer steps, by direct deposition of diamond films on a suitable substrate, as silicon nitride, silicon carbide and, the commercially more common, cemented carbide (WC-Co) with Co content lower than 6 wt.%. The thickness is generally in the range of 5 to 50 μm. Although this route is the simplest one, the problems with adhesion mainly on WC-Co substrates delayed the progress of the diamond direct coated tools. So, this kind of diamond cutting tool fabrication has been much investigated and is in continuous improvement. Some examples of the metallic and non-metallic materials along with their associated machining challenges that justify the use of the diamond tools are listed in Table 1.

Table 1. Examples of difficult-to-cut machine composite materials (adapted from [4])

Material	Example	Example of use	Machining issues
Hypereutectic silicon-aluminum alloy	A390 (18 wt.% Si particles in Al matrix)	Reduced-weight, wear resistant, temperature-resistant pistons	Hard silicon particles are extremely abrasive
Metal matrix composites (MMC)	Duralcan (20wt.% SiC in Al matrix)	Brake rotors, light-weight structures	Hard SiC ceramic particles are extremely abrasive
Cemented tungsten carbide	WC- 25wt%Co (sintered WC grain in Co based alloy matrix)	High-fracture-toughness wear parts, mould industries	WC grains are very hard and strong bonded. The Co metal binder can react with the carbon in diamond
Structural aerospace composites	Carbon-epoxy (high density carbon fibers in epoxy polymer)	Stiff and light-weight support structures for commercial aircraft, strong and light-weight sporting goods	Carbon fibers are extremely abrasive
Glass fiber reinforced polymers (GFRP)	G10 (highly compressed fiber glass in epoxy polymer matrix	Light-weight, insulating circuit boards, low-cost structural composites	Glass fiber induce abrasive tool wear; polymer can cause corrosive (chemical) tool wear
Graphite	ISO 88 polycrystalline graphite	Electrodes to electrodischarge machining process (EDM) in mould industries	Abrasive aggregates of polycrystalline graphite are formed during machining and wears the cutting edge parts

There is some divergence about the thin-film CVD diamond coating tool performance, which denotes the need for the improvement of this kind of tool. The tool life would span an order of magnitude in terms of cutting time, with some tools wearing at about the same rate as that of a PCD tool (PCD is the acronym of Polycrystalline Diamond, a sintered composite of HPHT diamond grains in a cobalt binder brazed onto a WC-Co substrate) [6]. Shen [6] tested thin film diamond coated indexable WC-Co and Si_3N_4 ceramic inserts from a large number of sources in dry machining of the hypereutectic A390 (18wt.% Si) aluminum alloy. He associated the different adhesion strengths with the reflected differences in the turning performance. Among the tools tested, two or three coatings sources were able to have good film-to-substrate adhesion and a machining performance comparable to that of the PCD inserts. When comparing the flank wear of these tools at the same cutting conditions, he found a great inconsistency even within a batch or among batches by the same coating source.

Uhlmann and co-workers [7] compared Si_3N_4 and WC-Co diamond coated tools in turning, milling and drilling of AlSi and AlCu alloys, and a fiber-reinforced polymer. They show that, in turning operations, the diamond coated silicon nitride tools provide the higher wear resistance, increasing the tool life and enlarging the usable cutting speed range. For milling, in some cases,

2

and in all cases for drilling, the diamond coated WC-Co tools presented the best performance because of the high thermo-mechanical stress imposed to the tools in such operations.

In a study carried out by Uhlmann and Brucher [8] with thin film diamonds and thick brazed films tools in machining of the same AlSi alloy, the authors arrived to different conclusions. They found that the thick brazed films could be successfully used with a tool life of 7 min (wear criteria adopted of VB=0.2mm), while it was not possible to conduct tool life tests on both WC-Co and Si_3N_4 ceramic coated tools due to the occurrence of film delamination after cutting times of only 30s. These results contradict the previous, above cited, work of Uhlmann et al., which reports values of 50 min of tool life adopting the same wear criteria. The turning parameters were similar concerning to cutting speed, but differ on feed and depth-of-cut conditions. In the first work, these parameters were: 0.04 and 0.8mm for feed and depth, respectively, while in the second they were: 0.1 and 0.5mm.

D'Errico and Calzavarini [5] reported the turning of metal-matrix composites (MMC) based on SiC (20wt.%) reinforced Al matrix (Duralcan) with CVD thick diamond brazed (~500 μm) and thin-film coated WC-Co (20-50 μm) from different sources and compared with PCD tools. They concluded that the thick film can be considered as a competitor for PCD by its superior wear resistance (binder-free, pure diamond nature), reducing the tendency for diamond grain "pullout" when eroded by the SiC particles. On the other hand, the thin coated tools failed by coating delamination after a very short cutting time (the average was 10 s). More recently, Chou and Liu [9] demonstrated, in turning of the same MMC material with diamond coated WC-Co tools, that adjusting the cutting parameters (mainly diminishing the feed), the film could delay the onset of tool failure by film delamination from few seconds until almost 15 minutes, although all the tested tools failed by this way before reaching a tool life criteria by abrasion. Another problem addressed in that work was the adhesion of the work material on the asperities of the diamond film, which forms a built-up-edge (BUE) formation on the rake face. This can be very harmful to diamond coatings, since it can also cause chipping at the cutting edge of the tool when the adhesive junctions are broken [6].

In resume, direct comparisons and conclusions seems to be very difficult to be assumed since the properties of both diamond coatings and substrates, as well as their manufacturing process (substrate characteristics, surface pretreatments, diamond deposition conditions) certainly differs from producer to producer, affecting the overall quality of the final product. In addition, other factors as cutting parameters, cutting conditions (lathe stability, use and type of lubricant) and workpiece characteristics (mechanical properties, homogeneity, dimensions) will direct affect the machining performance of a tool.

Nanocrystalline diamond

Friction between the tool/workpiece contact zones is influenced by the nature of the materials pairs, but also and in a great extent by the quality of the cutting edge, namely the tool surface roughness. A number of techniques were developed to polish the free surface of diamond films, as mechanical [10], thermo-mechanical [11], thermo-chemical [12] and laser [13]. But the stress imposed by some of these techniques as well as the time spent and the complexity of the equipments needed led to the development of diamond film growth with controllable grain texture and/or very small grain sizes, in order to diminish the inherent roughness created by columnar structure of the CVD growth [14].

3

To overcome this drawback, research efforts have started to focus on nanocrystalline diamond (NCD) coatings, due to its small grain size and very low surface smoothness [15-19]. One of the main advantages of these coatings is the almost constant crystallite size of the diamond through the entire film cross-section, contrarily to columnar growth observed in microcrystalline CVD diamond [15]. The control of CVD diamond film microstructure is achieved by changing the deposition parameters to growth or re-nucleation conditions of diamond. Such parameters are highly dependent on the technique used as well as geometric factors of the reactor chamber. NCD films can be produced in a microwave plasma CVD reactor from a variety of feed gas mixtures such as fullerenes/Ar, CH_4/Ar, CH_4/N_2, or CO/H_2 [19-22]. The diamond crystallite size typically varies from 3 to 30 nm and the intrinsic surface roughness from 15 to 40 nm [15]. The hot filament CVD technique can also be used to grow such coatings either by applying a bias current that can be used to enhance growth rate and minimize grain size and surface roughness [23] or by carefully adjusting deposition conditions under H_2/Ar/CH_4 gas mixtures [21,24,25]. The increase of CH_4/H_2 ratio also enhances the secondary diamond nucleation, but above a given ratio, graphite may form and prevents diamond nucleation [23].

NCD is normally described as nanocrystalline diamond grains embedded in a predominant tetrahedrally coordinated amorphous carbon network [15]. However, this kind of NCD film are often termed "cauliflower" or "ballas" diamond, because of the substantial amount of sp2-bonded nature of the grain boundaries [21,22]. The so-called "Ultra-Nanocrystalline" diamond (UNCD) is said to differ from NCD due to its much smaller grain sizes (3-5 nm) and have an abrupt grain boundaries with negligible sp2-bonded carbon [21].

Silicon nitride ceramics as substrate material for CVD diamond

When considering tribological and mechanical applications, adhesion of the diamond film to the substrate determines the success of the component in service. This was the reason for the fairly slow progress of the thin-film CVD diamond coated tools, mainly because cemented carbides, the most common substrate for cutting tools [26,27], provided unsatisfactory and inconsistent adhesion between the diamond coating and the substrate. Nevertheless, great progress on different methods of mechanical and chemical substrate surface pretreatments have been made [28,29].

The CVD process ideally requires a substrate material with a thermal expansion coefficient similar to that of diamond (~$1\times10^{-6}K^{-1}$) in order to reduce the thermal induced stresses developed on the cooling step. Cemented carbide (cemented carbide) possesses a higher thermal expansion coefficient (~$6\times10^{-6}K^{-1}$), leading to a higher thermal mismatch. A promising solution was proposed by others, including our group, and consists on the use of silicon nitride ceramic (~$2\times10^{-6}K^{-1}$) cutting substrates. Furthermore, these ceramics do not induce graphite formation at the interface during deposition and enhance chemical bonding.

Machining of cemented carbide by chip removal

Cemented carbide parts are used in a wide branch of industries, including chemical, medical, automotive, packaging, textiles, mining, oil and gas, siderurgy, and for shaping and drawing technologies. The machining of cemented carbide by chip removal using superhard cutting inserts is a recent technology, due to the high hardness and abrasive nature of cemented carbide. The option for turning process instead of grinding brings several advantages as: better

4

surface quality, reduction of production steps, reduction of product time delivery, shortening the manufacture cycles, higher geometry flexibility (corners, radius and grooves) and finally less energy consumption.

A tenfold decrease on the machining time compared with the conventional diamond wheel grinding method was achieved by thick CVD diamond brazed tools in facing WC-27wt%Co [30]. In such applications, polycrystalline diamond (PCD) and polycrystalline cubic boron nitride (PCBN) are the most established market options, together with thick CVD diamond brazed films. Just a few works are devoted to this issue [7,30-36], despite some of them are about micromachining, where the depth-of-cut is only of a few micrometers (3 μm) [33,34]. There is a technical guide on the use of PCD or PCBN tools associated to the binder phase content, typically cobalt, of the cemented carbide workpiece [32]. Accordingly, PCD is used when the binder content is below 18wt%, due to its superior abrasion resistance. However, PCBN should be used when the binder content is above this value, regarding its higher thermal and chemical stability. This is due to the chemical affinity between carbon and cobalt, and so, binder contents higher than 18wt% increases considerably the carbon solubility, leading to detrimental effects in its wear resistance properties [32].

The use of thin film CVD direct coated diamond tools in cemented carbide machining is a first example of a novel application of this material and it will be discussed further. It may configure an excellent alternative for PCBN and PCD tools, considering that CVD diamond higher hardness and absence of cobalt binder allows its use for machining a wider range of cemented carbide's cobalt content without need to have several types of cutting tools. Also, the technology of direct coating allows the manufacturing of diverse cutting insert shapes, including chip-breakers.

Bone drilling with CVD diamond tools

Dental implant surgery significantly moved forward in 1965 when Branemark introduced the first titanium dental implant [37]. For the implant threading, the bone must be previously drilled. Here, the main problem is overheating that may rise the bone temperature above the necrosis threshold range, 42°C to 47°C [38-41], resulting in the implant failure [42-46]. Materials with higher thermal conductivity and increased cutting efficiency compared to conventional metallic drills may be a very good alternative. Those are precisely excelling properties of CVD diamond, moreover if the drill body is made of Si3N4 ceramic to ensure adhesion of the coating to the tool substrate, as stated above.

Diamond tools are also biocompatible [47] and possess very high chemical stability, i.e., diamond wear debris, if present, are harmless when compared to metallic ones [47]. Moreover, CVD diamond in its nanocrystalline grade was demonstrated to exhibit a higher resistance to bacterial colonization than medical steel and titanium [48], avoiding bacterial infection from such foreign particles. The performance of NCD coated Si3N4 ceramic drills in bone drilling is a second novel application for CVD diamond tools that will be further discussed.

EXPERIMENTAL

Fully dense Si3N4 ceramic discs are produced by powder technology including pressureless sintering at 1750°C for 2 hours in N2 atmosphere [34]. For the cutting tests, the ceramic parts are ground to standard normalized geometries of round shaped indexable inserts, accordingly to ISO

5608 insert identification system. For bone drilling experiments, the Si_3N_4 ceramics are machined to the shape of a commercial steel (AISI 420) drill bit. Surface treatments before CVD diamond deposition included: flank face grinding with diamond wheel, rake face polishing with 15 µm diamond slurry, etching by CF4 plasma (1 h), and scratching/seeding in a diamond suspension in n-hexane by ultrasonification (1 h).

Diamond growth is conducted by hot filament chemical vapor deposition (HFCVD). For the cutting inserts, three types of diamond films are produced, which are labeled in Table 2 as: MCD, for microcrystalline diamond; NCD1, for nanocrystalline diamond type 1; and NCD2, for a finest nanometric grade. Film thickness is evaluated from SEM cross-section micrographs. The surface roughness is determined using AFM microscopy from 50 µm×50 µm scan areas. Diamond crystallite sizes are estimated by the broadening of the XRD diffraction peak at $2\theta\sim44°$, corresponding to the diamond (111) plane. In the case of MCD films, the crystals average size is calculated from SEM views.

In the case of the Si_3N_4 drill bits, these are coated with a bi-layer of MCD/NCD2, following the respective deposition parameters, except the filament temperature that is lowered to 2050°C due to the vertical alignment around the drill. The final thickness is about 1.5 µm and 2.5 µm, respectively for the MCD layer and the NCD2 one.

Table 2. Hot filament deposition parameters and CVD diamond characteristics.

Tool Type	CH_4/H_2	Ar/H_2	Gas flow (sccm)	Total pressure (mbar)	Filament temperature (°C)	Substrate temperature (°C)	Growth rate ($\mu m \cdot h^{-1}$)	Thickness (µm)	RMS (µm)	Grain size
MCD	0.02	-	100	25	2300	850	2.7	42	0.78	12µm
NCD1	0.03	-	100	25	2300	850	3.1	50	0.23	43nm
NCD2	0.04	0.1	50	50	2300	750	1.3	23	0.26	27nm

SEM cross-sections of the three CVD diamond grades used for the cutting inserts coating are given in Figure 1. The respective insets show the respective top view morphologies. The MCD film has a columnar structure, originating large diamond grains at the free surface (Figure 1a). On the contrary, the nanocrystalline NCD1 and NCD2 films present a very flat surface. The relatively high RMS surface roughnesses given in Table 1 are mainly the result of the nominal RMS value of the Si_3N_4 substrate, which is about 0.4 µm. Good surface finishing can be achieved by the lower diamond crystallites sizes, NCD1 and NCD2, reaching the accepted industrial surface roughness parameter of about 0.2 µm.

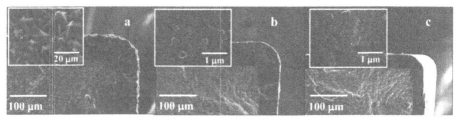

Figure 1. SEM cross-sections and top view insets of MCD(a); NCD1 (b) and NCD2 (c) tools. [36]

The turning tests are done using a CNC lathe equipped with a three-axis piezoelectric dynamometer platform. The workpiece surface quality is determined using a portable profilometer. The turning tests are conducted under dry cutting conditions. The workpiece material is a WC-25 wt.% Co cylinder (∅=32 mm, length=60 mm). The cutting performance of the three types of diamond coated tools is evaluated at fixed conditions: speed=15 m min^{-1}, depth-of-cut=0.1 mm and feed=0.1 mm rev^{-1}. The resultant wear modes are measured accordingly with the ISO 3685 standard using optical and SEM microscopes.

The drilling tests are performed in a universal mechanical testing machine equipped with a variable speed drill. The human mandible bone is simulated using a laminated test block formed by a solid rigid polyurethane foam (trabecular bone) with a 2mm thick upperlayer of E-Glass-filled epoxy sheet (cortical bone). Spindle speed varied in the range of 50 to 1400 rpm (cutting speed of 0.5 to 13.2 m/min) and the infeed rate between 7.5 to 30 mm/min. Local temperature is monitored by two thermocouples placed in the polymer at different levels (TC1 placed immediately after the epoxy sheet at 3 mm depth, and TC2 is placed at the end of drilling hole at 15 mm).

RESULTS AND DISCUSSION

Machining of cemented carbide with micro and nanocrystalline CVD diamond

The force components during cemented carbide turning are given in Figure 2a: depth-of-cut (F_d), main (F_c) and feed (F_f) forces. The surface roughness of the MCD coated insert leads to excessive friction between the tool and the workpiece materials and thus high forces are generated during the cutting operation. On the contrary, cutting forces for both the two NCD grades are very similar, being relatively low. As the inserts are circular shaped, the contact area with the workpiece is high and thus the F_d component is the highest in all grades.

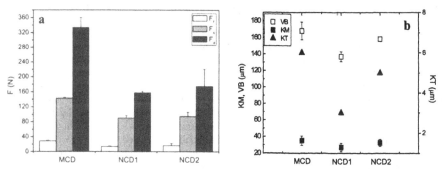

Figure 2. Cutting forces (a) and wear values at rake and flank faces (b) of the distinct diamond grades after one turning pass (~60 m cutting length, 4 min of cutting time).

Figure 2b summarizes the set of values of the crater centre distance (KM), average flank wear width (VB) and crater depth (KT). The lowest values occur for the NCD1 grade and the largest ones for the MCD coated inserts. For all grades, flank wear is the predominant kind of tool wear, as VB values denote. According to the ISO Standard 3685, a VB value of 300 µm is adopted for tool life criterion. This value of VB limit is not achieved after 4 min of cutting time by any of the tools. A remarkable feature of all the turning experiments is the absence of film delamination demonstrating that the adhesion strength to the Si_3N_4 substrates is adequate.

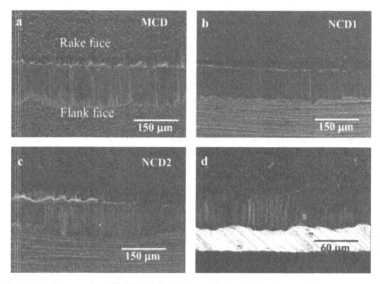

Figure 3. SEM micrographs of rake and flank faces of the distinct CVD diamond tools. (a) to (c) correspond to MCD, NCD1 and NCD2 coatings, respectively. Micrograph (d) is a top view of the MCD cutting tool after three machining passes (12 min of cutting time) [36].

The dominant wear mechanism in cemented carbide cutting by diamond coated tools is abrasion caused by loose diamond debris and WC hard particles from the sliding action of the chip and workpiece rubbing on the rake and flank faces, respectively [35]. This can be seen in Figure 3, in general. In the MCD insert, the higher surface roughness leads to a higher probability of diamond microchipping and transgranular fracture, explaining the higher wear values. Nevertheless, the MCD tool is tested for longer times, reaching the VB limit after 12 min (Figure 3d). The different behavior of the two NCD grades can be explained by the higher amount of adherent material in NCD2 tool, that is characterized by a relatively higher surface roughness (Table 2), as a consequence of the nanocrystalline clustered morphology (Figure 1c). An further advantage of the NCD tools comparing to the MCD one is the better workpiece surface finishing being kept below the industrial accepted level of Ra=0.2 μm.

Bone drilling behavior of bi-layered CVD diamond coated Si_3N_4 bits

Figure 4a shows typical plots of the feed force evolution during the drilling tests. For each drill type (steel and CVD diamond bi-layer coated Si_3N_4), two different curves are given, corresponding to the maximum spindle speed (1400 rpm) and the minimum admissible spindle speed for the steel drill (350 rpm). In all cases, the feed force increases to a maximum when the drill tip is still cutting the denser upper layer of the polymer (that mimics the cortical bone) [44]. After that, a new cutting regime takes place, with nearly steady low force values when the mild polymeric foam (that emulates the trabecular bone) is drilled.

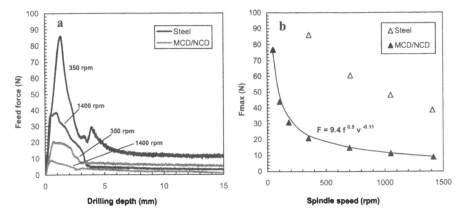

Figure 4. Feed force evolution as a function of the drilling depth (a) and maximum feed force values [49].

Although the force evolution in both drill types is similar, the values attained by the bi-layered CVD diamond coated drills are remarkably lower than those of the steel drill bit. This is the result of the more abrasive action of the diamond coated tool that sums to the shearing action of the tool edge leading to a lower cutting effort. The spindle speed has a net effect on the drilling action by decreasing the maximal feed force required for the task (Figure 4b). Again, the diamond tool is much better, for which an expression on the dependence of feed (f) and spindle speed is given in Figure 4b, denoting the prevailing importance of feed. Of noteworthy importance is the potential of the MCD/NCD2 coated tool to cut at very low spindle speeds (100-180 rpm), contrarily to the commercial steel drill. Bearing in mind that a threshold of 45N [38] is considered to be the uppermost value for normal loading in odontological drilling, the steel drill cannot operate below 1200 rpm in the absence of liquid cooling, while the diamond coated tool can be used from 100 rpm onwards. This ability reduces local heating, and thus bone damaging, while it allows collecting of bone debris for self-regeneration purposes.

The local polymer temperature during drilling is represented in Figures 5a and 5b, respectively for the steel drill and the diamond coated one. The temperature evolution on thermocouples TC1 and TC2 is firstly plotted as a function of the drilling depth, until removal at the end of cut (15 mm), and from that instant as a function of time. Taking the data from the TC1 thermocouple, four regions (I-IV) can be identified: I, when no variation of the initial temperature is detected until the drill crosses the TC1 placement; II, characterized by steep heating which comes from the locally generated heat at the contact between the drill tip and the polymer; III, when the increment of the distance between the heat source (the drill tip) and the TC1 positioning exceeds the temperature rising on the drill and on the polymer by heat accumulation; IV, corresponding to the polymer natural cooling after the drill removal. Concerning the TC2 thermocouple placed near the hole end at 15 mm, heating starts to be detected before the drill tip approaches this level; after the drill removal, this thermocouple still registers an increase in temperature as a result of the polymer thermal inertia.

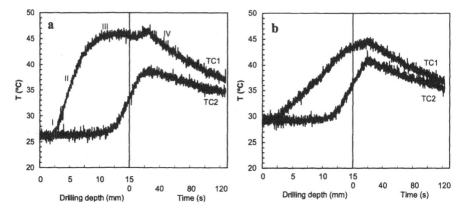

Figure 5. Temperature evolution as a function of the drilling depth and of the time after the drill bit removal, acquired by thermocouples TC1 and TC2, in drilling at a spindle speed of 1050 rpm with the steel drill bit (a). The same but for the bi-layered CVD diamond coated drill bit (b). [49]

The temperature evolution on the MCD/NCD2 coated Si_3N_4 drill test shows a net difference on the TC1 thermocouple data, Figure 5b, comparing to the steel tool, Figure 5a: the temperature gradient is less steep, the maximum temperature is lower, with a reduction of about 4°C. These are expected results taking into account that the drilling forces are much lower as before discussed. For the diamond coated drill the maximum temperature is always below 45°C at of 1050 rpm of spindle speed, below the threshold temperature for heat induced bone injury (47°C for 1 minute [39]).

Both types of drill bits are observed by SEM after testing [49]: the cutting edge of the commercial steel drill becomes chamfered as a result of the wear process, leading to higher drilling forces and excessive heating; on the contrary, the bi-layered MCD/NCD2 coated drill bit presented no signs of delamination or edge deterioration keeping intact the initial roundness of the coated edge.

CONCLUSIONS

Silicon nitride inserts are successfully coated by CVD hot filament method with continuous and highly adherent diamond films of nano- (27 and 43 nm) and conventional 12 μm micrometric grain size types. The 43nm nanocrystalline grade presented the best behavior in machining of WC–25wt.% Co cemented carbide regarding cutting forces, tool wear and workpiece surface finishing. The conventional micrometric grade, featured by the presence of high asperities originated by the large crystals, suffered a higher abrasive action from the loose hard particles carried by the chip flow and from rubbing at the contact with the workpiece.

A dense and homogeneous bi-layer of microcrystalline/nanocrystalline diamond is also successfully deposited by HFCVD on Si_3N_4 drill bits for odontological purposes. After drilling experiments on a laminated block that simulating the human mandible bone, no signs of delamination or wear are found on the diamond coating attesting the good adhesion to the substrate. The CVD diamond coated drill bit presents a superior behavior in comparison to a conventional steel drill bit regarding: i) the maximum force applied in the drilling procedure that is approximately four times lower; ii) the maximum temperature obtained for the coated drill bits that is smaller in about 4 °C, only reaching the maximum limit of 47 °C for the highest spindle speed (1400 rpm); iii) the potential for drilling at low spindle speeds (100 rpm), thus producing less heat and reducing the bone damage that allows collecting of bone debris for self-regeneration purposes; iv) the non-toxicity of the wear particles.

ACKNOWLEDGMENTS

F. Almeida, M. Amaral and E. Salgueiredo acknowledge FCT for grants SFRH/BPD /34869/2007, SFRH/BPD/26787/2006 and SFRH/BD/41757/2007, respectively.

REFERENCES

1. J. C. Angus, in *Synthetic Diamond: Emerging CVD Science and Technology*, edited by K. E. Spear and J. P. Dismukes (John Wiley & Sons, New York, 1994), cap. 2.
2. P. W. May, *Royal Soc. Phil. Trans. A.* **358**, 473 (2000).
3. J. C. Angus, H. A. Will and W. S. Stanko, *J. Appl. Phys.* **39**, 2915 (1968).
4. B. L. Cline and J. M. Olson, in *Diamond Films Handbook*, edited by J. Asmussen and D. K. Reinhard (Marcell Dekker, New York, 2001).
5. G. E. D'Errico and R. Calzavarini, *J. Mater. Process. Technol.* **119**, 257 (2001).
6. C. H. Shen,. *Surf. Coat. Technol.* **86-87**, 672 (1996).
7. E. Uhlmann, U. Lachmund and M. Brücher, *Surf. Coat. Technol.* **131**, 395 (2000).
8. E. Uhlmann E and M. Brücher, *Annals of CIRP* **51**, 49 (2002).
9. Y. K. Chou and J. Liu, *Surf. Coat. Technol.* **200**, 1872 (2005).
10. W. J. Zong, D. Li, T. Sun and K. Cheng K, *Diam. Relat. Mater.* **15**, 1424 (2006).
11. S. Jin, L. H. Chen, M. McCormack and M. E. Reiss, *Appl. Phys. Lett.* **63**, 622 (1993).
12. B. Bhushan, V. V. Subramaniam, A. Malshe, B. K. Gupta and J. Ruan, *J. Appl. Phys.* **74**, 4174 (1993).
13. A. M. Ozkan, A. P. Malshe and W. D. Brown, *Diam. Relat. Mater.* **6**, 1789 (1997).
14. A. Erdemir, G. R. Fenske, A. R. Krauss, D. M. Gruen, T. McCauley and R. T. Csencsits, *Surf. Coat. Techonol.* **120-121**, 565 (1999).
15. N. Toprani, S. A. Catledge and Y. K. Vohra, *J. Mater. Res.* **15**, 1052 (2000).
16. A. C. Ferrari and J. Robertson, *Phys. Rev. B* **63**, 121405 (2001).
17. S. Bhattacharyya, O. Auciello, J. Birrel, J. A. Carlisle, L. A. Curtiss, N. A. Goyette, D. M. Gruen, A. R. Krauss, J. Schlueter, A. Sumant and P. Zapol, *Appl. Phys. Lett.* **79**, 1441 (2001).
18. S. Gupta, B. R. Weiner and G. Morell, *J. Mater. Res.* **17**, 1820 (2002).
19. S. A. Catledge, J. Borham, Y. K. Vohra, W. R. Lacefield and J. E. Lemons, *J. Appl. Phys.* **91**, 5347 (2002).
20. P. Bruno, F. Bénédic, A. Tallaire, F. Silva, F. J. Oliveira, M. Amaral, A. J. S. Fernandes, G. Cicala and R. F. Silva, *Diam. Relat. Mater.* **14**, 432 (2005).
21. P. W. May, J. A. Smith and Y. A. Mankelevich, *Diamond Relat. Mater.* **15**, 345 (2006).
22. R. Haubner and B. Lux, *Int. J. Refract. Met. Hard Mater.* **20**, 93 (2002).

23. M. Vojs, M. Veselý, R. Redhammer, J. Janík, M. Kadleciková, T. Danis, M. Marton, M. Michalka and P. Sutta, *Diam. Relat. Mater.* **14**, 613 (2005).
24. M. Amaral, A. J. S. Fernandes, M. Vila, F. J. Oliveira and R. F. Silva, *Diam. Relat. Mater.* **15**, 1822 (2006).
25. Y. F. Zhang, F. Zhang, Q. J. Gao, X. F. Peng, and Z. D. Lin, *Diam. Relat. Mater.* **10**, 1523 (2001).
26. R. Polini, M. Barletta and M. Delogu, *Thin Solid Films* **515**, 87 (2006).
27. M. Amaral, F. J. Oliveira, M. Belmonte, A. J. S. Fernandes, F. M. Costa and R. F. Silva, *Surf Eng.* **19**, 410 (2003).
28. K. Mallika and R. Komanduri, *Wear* **224**, 245 (1999).
29. R. Polini, *Thin Solid Films* **515**, 4 (2006).
30. M. Belmonte, P. Ferrro, A. J. S. Fernandes, F. M. Costa, J. Sacramento and R.F. Silva, *Diam. Relat. Mater.* **12**, 738 (2003).
31. M. Belmonte, F. J. Oliveira, J. Sacramento, A. J. S.Fernandes and R. F. Silva, *Diam. Relat. Mater.* **14**, 843 (2004).
32. M. Collier and J. Cheynet, *Proceedings of PM2002 Hard Materials and Diamond Tooling Congress and Exhibition* (7–9th October 2002, Lausanne, Switzerland, 2002).
33. Y. C. Yen, A. Jain and T. Altan, *J. Mater. Process. Technol.* **146**, 72 (2004).
34. F. A. Almeida, F. J. Oliveira, M. Sousa, A. J. S. Fernandes, J. Sacramento and R.F. Silva, *Diam. Relat. Mater.* **14**, 651 (2005).
35. F. A. Almeida, A. J. S. Fernandes, R.F. Silva and F. J. Oliveira, *Surf. Coat. Technol.* **201**, 1776 (2006).
36. F. A. Almeida, M. Amaral, F.J. Oliveira and R.F. Silva, *Diam. Relat. Mater.* **15**, 2029 (2006).
37. A. N. Cranin and J. E. Lemons, *Biomaterials Science - An Introduction to Materials in Medicine*, (Elsevier Academic Press, 2004).
38. G. E. Chacon, D. L Bower, P E. Larsen, E. A. McGlumphy and F. M. Beck, *Int. J. Oral Maxillofac. Implants* **64**, 265 (2006).
39. R. Eriksson and T. Albrektsson., *J. Prosth. Dent.* **50**, 101 (1983).
40. A. Eriksson and T. Albrektsson, *Int. J. Oral Surg.* **1**, 115 (1982).
41. R. Eriksson, T. Albrektsson and B. Magnusson., *Scand. J. Plastic Reconst. Surg.* **18**, 261 (1984).
42. M. Sharawy, C. E. Misch, N. Weller, S. Tehemar, *J. Oral Maxillofac. Surg.* **60**, 1160 (2002).
43. R. A Eriksson and R. Adell., *J. Oral Maxillofac. Surg.* **44**, 4 (1986).
44. S. Karmani, Current Orthopaedics **20**, 52 (2006).
45. W. Allan, E. D. Wiliams and C. J. Kerawala, *Br. J. Oral Maxillofac. Surg.* **43**, 314 (2005).
46. T. Udiljak, D. Ciglar and S. Skoric, *Adv. Prod. Eng. Manag.* **2**, 103 (2007).
47. P. Aspenberg, A. Anttila, Y. T. Konttinen, R. Lappalainen, S. B. Goodman, L. Nordsletten and S. Santavirta, *Biomaterials* **17**, 807 (1996).
48. W. Jakubowski, G. Bartosz, P. N. Szymanski and B. Walkowiak, *Diam. Relat. Mater.* **13**, 1761 (2004).
49. E. Salgueiredo, F. A. Almeida, M. Amaral, A. J. S. Fernandes, F. M.Costa, R. F. Silva and F. J. Oliveira, *Diam. Relat. Mater.* **18**, 264 (2009).

Mater. Res. Soc. Symp. Proc. Vol. 1243 © 2010 Materials Research Society

Fabrication of Porous Glass by Hot Isostatic Pressing

Koji Matsumaru, Ryoichi Hanawa and Kozo Ishizaki
Nagaoka University of Technology,
Nagaoka, Niigata 940-2188, JAPAN

ABSTRACT

In this study, a fabrication method for porous glass is established by using hot isostatic pressing (HIPing). Pyrex glass powders and argon are used as raw material and applied gas, respectively. HIPing pressures are in the range 7–150 MPa and 1100 °C as HIPing temperature, this pressure is applied before heating the sample. The holding period at HIPing temperature is 1 h. The HIPed sample is heat treated for 10 min of holding time at a temperature of 700-1000 °C. The HIPed samples are expanded by heat treatment. The porosity of heat treated samples increases with the increase of HIPing pressure. Open pore porosity increases with increase of a heat treatment temperature. The proposed method uses the high solubility of gases in molten glass and gives control of porosity and structure of pores by adjusting HIPing pressure and heat treatment temperature.

INTRODUCTION

Porous materials have many different applications. Specialty, porous ceramics and glass have attracted attention due to its variation of physical and mechanical properties. Thermal or sound insulators, light structure materials, filters, catalyst beds and grinding wheels are typical applications of porous materials [1]. Optimal pore size and porosity are required for each application [1]. Therefore, porous materials are produced through many different methods to control pores. Sintering is a typical method to obtain porous materials. Conventionally, porous materials are produced by using short time or low temperature sintering. To increase open porosity, pore forming agents are mixed with ceramics powders. Pore forming agents are evaporated or burned out during sintering, and as a result pores are formed.

Foam glasses are used as insulators in houses or as raw materials for roads. Hara *et al.*, have fabricated foam glass with zeolites for water purification since it absorbs ammonia or heavy metals from river waters. Calcium carbonate is also used to produce foam glass [2]. Kunijima *et al.*, have proposed foam glass as absorber of electromagnetic waves since these materials have conducting whiskers. Additionally, the foam glass is also fabricated by using pore foaming agents [3]. Tani *et al.* have fabricated grinding wheels by using foam glass with good results for silicon wafer grinding [4]. In all of these investigations, foam glass is required as a foaming agent.

In this research, a new fabrication method is proposed by using the high solubility of gases under a high pressure and the viscosity properties of glass at high temperature. In this method, inert gas dissolves in molten glass under high pressure. Then, a foam glass is fabricated by using the solubility gap under a pressure difference and the low viscosity of glass at the heat treatment temperature. In this manner, the effects of pressure (applied by Hot Isostatic Pressure, HIP) and heat treatment temperature on porosity and pore structure are investigated.

EXPERIMENTAL PROCEDURE

Pyrex glass powder is used as a raw material. Table 1 shows the investigated compositions of pyrex glass with an average powder particle diameter of 19 μm. The processing starts by adding 5 g of glass powder in a graphite crucible (size is 38 mm in diameter and 44 mm in height). This crucible is placed in the HIPing equipment (O_2-Dr. HIP, Kobe Steel Co., Ltd.) and subjected to the cycle shown in Fig. 1. Argon is used in the HIP processing. The HIPing pressure is increased to its selected holding value before increasing the temperature. The holding pressures in use are 7, 70 and 150 MPa. The heating rate for all cases is 400 K/min. The holding time is 1 h at 1100 °C. HIPed samples are heat treated under atmospheric condition. The heat treatment temperatures are 700, 800 and 1000 °C. The heating rate is 300 K/min. The holding time is 10 min in all cases. Porosity is measured by the Archimedes method. Pores of samples are also observed by using a scanning electronic microscope.

Table 1 Compositions of pyrex glass

Compositions	SiO_2	Al_2O_3	B_2O_3	Na_2O	K_2O
Weight %	80	2	13	4	1

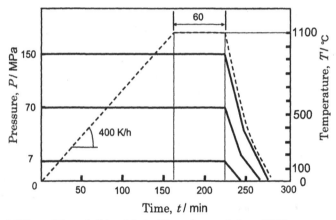

Figure 1 HIP conditions. Solid and dashed lines show variation of HIPing pressure and temperature, respectively.

RESULTS

Figure 2 shows a representative HIPed samples. Some sections of the sample HIPed under a pressure of 7 MPa show a dark gray color. On the other hand, higher HIP pressures produce samples with a lighter color. However the sample subjected to a HIP pressure of 150 MPa is considerably brighter than that produced under 70 MPa.

Ishizaki has reported phase diagrams under high total gas pressures [5]. I has been shown that as the pressure is increased the triple point shifts towards higher temperatures for reactions

14

such as: $C + O_2 = CO_2$, $2C + O_2 = 2CO$ and $2CO + O_2 = 2CO_2$. It is assumed that CO_2 is unstable in our experimental conditions since there is a high partial pressure of oxygen and then the stable substance is CO. Therefore, C and CO are stable for the experiments conducted under 7 MPa of HIPing pressure at 1100 °C. On the other hand, C is the stable substance under 70 and 150 MPa of HIPing pressure at 1100 °C. The stability of CO for experiments under lower HIPing pressure produces a gray color in the final molten glass i.e., there is contamination from the graphite crucible.

Figure 2. Photograph of HIPed samples. Sample color changes from gray to white with increase of HIPing pressure.

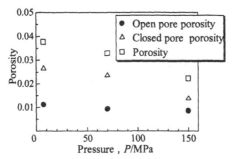

Figure 3 Porosity which is volume fraction of HIPed sample as a function of HIPing pressure. Porosity decreases with the increase of HIPing pressure.

Figure 3 shows the porositiy of HIPed samples as a function of applied pressure. All samples have a relatively low porosity. It decreases with the increase of HIPing pressure in all cases.

Figure 4 shows a photograph with a representative view of heat treated samples. All samples are heat treated at 1000 °C as above described. All samples expand by argon which dissolves under a high pressure and induces growth of closed pores with a high pressure as a consequence of heat treatment. All samples become a foam glass without a transformation.

15

HIPed samples under pressures of 7 and 70 MPa retain a circular shape. On the other hand, the sample HIPed under a pressure of 150 MPa, apparently explodes during heat treatment.

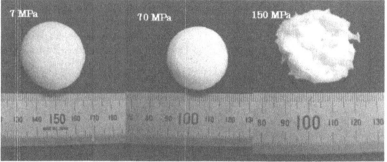

Figure 4. Photograph of heat treated samples at 1000 °C. The scale is given in cm.

Figure 5 shows the porosity of samples after heat treatment at 1000 °C. The porosity shows almost no variation for samples HIPed under 7 and 70 MPa of pressure. On the other hand, the sample produced under the highest HIP conditions (150 MPa) develops the highest porosity with a large fraction of closed pores. This is most likely due to the increase of the Ar solubility in glass as the HIPing pressure is increased.

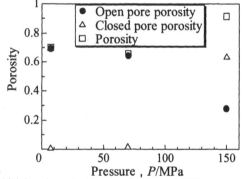

Figure 5. Porosity which is volume fraction as a function of HIPing pressure for samples after heat treatment at 1000 °C.

Figure 6 shows cross sectional views of the samples after thermal treatment at 1000 °C. The pore size varies for all samples apparently without connection to the HIPing pressure, the pore size is lowest for a processing pressure of 70 MPa. Figure 7 shows a pore growth process during a heat treatment. Fig. 7a, Fig. 7b and Fig. 7c correspond to the HIPing pressure of 7, 70 and 150 MPa, respectively. Since the solubility of gases increases with an increase of HIPing pressure, then the number of pores increases proportionally with the increase of HIPing pressure. Additionally, a high pressure inside of closed pore creates pore expansion. On the other hand, open pores are produced by connecting closed pores. Open pores can not hold a high pressure gas, and can not expand. Pore size increases as the closed pores expand, and ceases to increase as the closed pores become open pores by connecting them.

Figure 6 Cross section views of samples after heat treatment at 1000 °C.

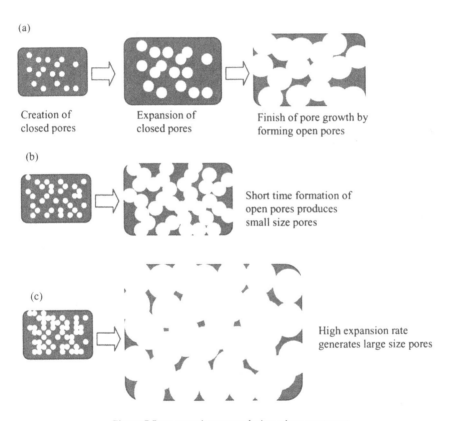

Figure 7 Pore growth process during a heat treatment.

Figure 8 shows the porosity as a function of heat treatment temperature for samples produced under a HIPing pressure of 70 MPa. All samples have similar values of porosity with a light increase of the volume fraction of open pores as a function of treatment temperature.

Closed pores connect each other and become open pores during heat treatment. The proposed method for foam glass gives control of a ratio of open pore porosity to closed pore porosity by a heat treatment temperature, because viscosity of glass decreases with the increase of a heat treatment temperature. Therefore, proposed method is expected to be a new process for control of properties of foam glass due to properties for foam glass can be changed by the ratio.

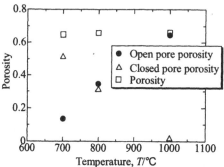

Figure 8 Porositiy which is volume fraction of hear treated sample of HIPing pressure at 70 MPa as a function of a heat treatment temperature.

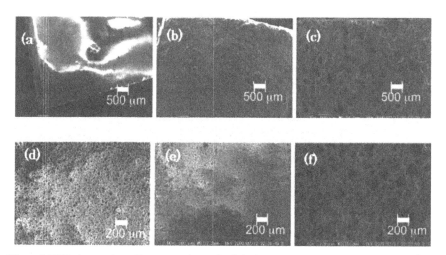

Figure 9 SEM observation of heat treated samples. The heat treatment temperatures are 700 °C for (a,d), 800 °C for (b,e) and 900 °C (c,f). The upper image corresponds to a surface view while the lower image shows a cross section of the samples in all cases.

Figure 9 shows SEM observations of heat treated samples. Figs. 9a,d, Fig. 9b,e and Figs. 89c,f correspond to a temperature of 700, 800 and 900 °C, respectively. The upper image corresponds to a surface view while the lower image shows a cross section of the samples in all cases. Open pores are apparent on the surface of all heat treated with an increase in number as a function of the heat treatment temperature. Pore size in heat treated samples also increases for higher heat treatment temperatures. The viscosity of glass decreases as the heat treatment temperature is raised. Therefore, closed pores grow with an increase of temperature and connect among themselves more effectively until the porosity consists almost exclusively of open pores.

CONCLUSIONS

A method to fabricate porous glass by using HIP is explored. This method uses only the solubility of gases in molten glass under a high pressure of inert Ar gas. Porosity is controllable by HIPing pressure. Structure of pores is changed by heat treatment temperature.

ACKNOWLEDGMENTS

The authors wish to express their gratitude to the Japanese government for partially supporting this work through the Promotion of Independent Research Environment for Young researchers of the Ministry of Education, Culture, Sports, Sciences and Technology.

REFERENCES

1. K. Ishizaki, S. Komarneni and M. Nanko,"Porous Materials: Process Technology and Applications", Kluwer Academic Publishers, Dordrecht, the Netherlands, 1998
2. Y. Hara, H. Matsuo and K. Ochiai, Japan patent, P2009-40623A
3. T. Kunisima, T. Iwao, M. Toyoda, H. Mabuchi and M. Yoshida, Japan patent, P2004-179354A
4. Y. Tani, N. Okumura and Y. Kamimura, "Development of high-porosity silicate wheel utilizing a foam water glass", The Japan Society for Abrasive Technology, Vol. 49, No. 7, (2005), pp. 386-390
5. K. Ishizaki, "Phase Diagrams under High Total Gas Pressures – Ellingham Diagrams for Hot Isostatic Press Processes", Acta Metal, Mater, Vol. 38, No. 11, (1990), pp. 2059-2066

Mater. Res. Soc. Symp. Proc. Vol. 1243 © 2010 Materials Research Society

High-Temperature Oxidation Resistant Nanocoatings on Austenitic Stainless Steels.

Hugo F. Lopez
Materials Department
University of Wisconsin-Milwaukee
P.O. Box 784, Milwaukee WI 53209

ABSTRACT

In recent years, the increasing energy costs have lead to power utility industries to seek/develop high efficiency systems of production and of energy utilization. In addition, environmental concerns regarding greenhouse gas emissions are playing a major role in the development of clean energy systems. The development of metallic materials that can withstand elevated temperatures is among the viable alternatives to increase energy efficiency. Nevertheless, for this to happen, the corrosion and oxidation resistance of Fe- and Ni-based alloys needs to be significantly improved. Among the possible ways to enhance the life of high temperature alloys is the application of protective ceramic coatings. Conventional coatings are expensive and the protective effects controversial at times. An alternative which offers a great potential is the application of nano-ceramic coatings. Hence, in this work nanocrystalline coatings based on nano-CeO_2 are applied to an austenitic stainless steel 304L and then exposed to elevated temperatures. Weight changes are monitored as a function of time and the results are compared with uncoated alloys tested under similar conditions. In addition, computer simulations of possible rate limiting diffusion mechanisms are carried out. It is found that the nanocoatings provided remarkable high-temperature oxidation resistance and improved scale adhesion. In particular, it is found that the smaller the nanoparticles are, the more effective the nanocoatings in providing oxidation resistance.

INTRODUCTION

Environmental concerns related to the greenhouse gas emissions and to the energy crisis and costs have been forcing power utility industries into improving the efficiency of energy production. Among the viable alternatives to increase energy efficiency is the likelihood of safe operation at elevated temperatures. However, for this to take place it is necessary to improve the corrosion and oxidation resistance of high temperature Fe- and Ni-based alloys. These alloys can be used under extreme temperatures and environments through the use of surface coatings. Typically, the development of protective Cr_2O_3, Al_2O_3 and rare-earth-oxide scales prevents/reduces the high-temperature degradation of the alloy substrates. However, at elevated temperatures beyond 800°C the oxide scales developed are not able to provide the desired protection. Hence high temperature resistant coatings are employed [1-6]. Yet, a number of the available coatings are expensive and do not provide the desired protection. In recent years, the use of nano-structured materials in a wide variety of applications has attracted significant attention due to their unique physical and chemical properties. Among these applications, the use of nanocoatings for corrosion and high temperature protection of engineering alloys and steels can be found. Accordingly, in this work, the use of nano-ceria based coatings on the oxidation resistance of a 304 stainless steel is investigated. In addition, mechanisms are proposed for the exhibited oxidation behavior based on microscopic and macroscopic computer simulations of possible mechanisms.

EXPERIMENTAL PROCEDURE

Table I shows the chemical composition of the 304L stainless steel. Test coupons are cut from stainless steel sheets. The coupons are grinded using 600 and 1200 grit SiC paper and this is followed by ultrasonic cleaning in isopropanol. Nanocrystalline CeO_2 particles are purchased from Aldrich Chemicals Company, Inc. In addition, nanocrystalline CeO_2 is synthesized using the micro-emulsion method published by Wu [7] and Patil [8]. The solution containing nanocrystalline CeO_2 is extracted from the upper layer of the micro-emulsion system and collected in order to use it as a coating. Alternatively, the purchased CeO_2 is dispersed in ethanol-water solutions in order to achieve a uniform distribution during the coating of the stainless steel. Nano-particle size is measured using a light scattering technique based on the Malvern Instrument Zetasizer Nano ZS and the results are given in Table II.

Table I. Composition of 304L Stainless Steel (Wt %)

Alloy	Fe	C	Cr	Ni	Mn	Si	P	S	Mo	N	Cu
304L	Bal.	0.017	18.41	8.32	1.85	0.34	0.03	0.012	0.42	0.06	0.48

Table II. CeO_2 Nanoparticle Sizes

Nanoparticle	Nominal Size (nm)	Measured Size (nm)
CeO_2 (Synthesized)	3-5	3.45
CeO_2 (Purchased)	<20	97% = 14.6 3% = 329

The stainless steel coupons are manually dip coated in the prepared solutions. The coating process is repeated several times with intermediate drying at 200°C. The coating times are the same for all the coupons in order to obtain similar coating thicknesses. Isothermal oxidation tests are carried out on both bare and coated coupons in dry air at 800°C and 1000°C and the exhibited oxidation rates are determined. In addition, the exhibited oxide scale morphology is characterized by EDX and SEM backscattered electrons.

RESULTS AND DISCUSSION

Figure 1 shows the oxidation mass gain ($\Delta W/A$) versus time at 800°C and 1000°C for uncoated and coated 304L stainless steels. Notice from this figure that nano-crystalline cerium oxide coatings are highly effective in reducing the oxidation rates of the stainless steels. It is found that after 168 hours at this temperature, spallation of the scale on the uncoated steel is clearly evident. In contrast, the oxide scale in the coated sample remained highly adherent and intact even after over 300 hours. In addition, it is found that commercial cerium oxide nanocoatings are not as effective in reducing the oxidation rates when compared with the synthesized cerium oxide nano-coatings.

Similar trends are found at a 1000°C (see Fig 1b). Table III shows the exhibited magnitudes of the parabolic rate constant, kp^2. Notice that at 800°C the parabolic rate constant is reduced by over two orders of magnitude when synthesized CeO_2 is used as coating and by over an order of magnitude when CeO_2 is employed.

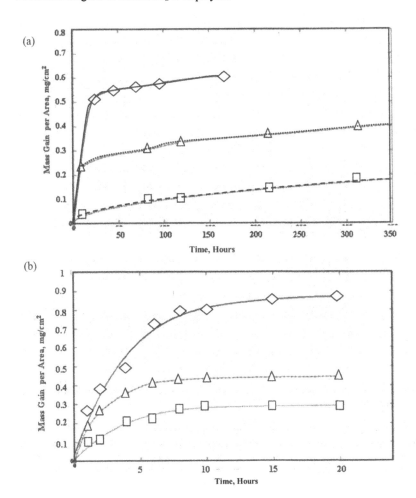

Figure 1. Mass gain per area vs. time plots of oxidation for uncoated and nanocrystalline cerium oxide coated grade 304L stainless steel in dry air at (a) 800°C and (b) 1000°C. —◇— Uncoated, spalling failure after 163 h at 800 C, – -□- – · CeO_2 (purchased) coated, ·····△····· CeO_2 (synthesized) coated.

23

Figure 2 shows the effect of nanocrystalline cerium oxide coatings on 304L stainless steels. Notice that when the coupons are oxidized at 1000°C for 10 hours, the uncoated sample shows significant spallation. In contrast, the coated one is uniformly oxidized, with no spallation. The apparent spallation on the lower-left corner of this sample is not spallation but increased protection due to a thicker CeO_2 at this location.

Table III. Exhibited Parabolic Rate Constants.

Temperature (°C)	Parabolic Rate Constant ($mg^2 cm^{-4} h^{-1}$)		
	Uncoated	Coated with Synthesized Oxide	Coated with Purchased Oxide
800	8.7×10^{-3}	9.6×10^{-5}	4.1×10^{-4}
1000	4.0×10^{-2}	4.4×10^{-3}	9.3×10^{-3}

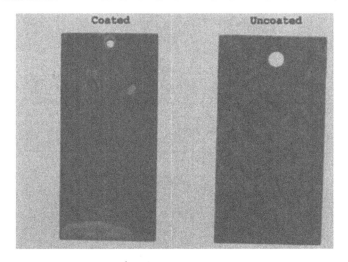

Figure 2. Uncoated and synthesized nanocrystalline cerium oxide coated 304 stainless steel after heating at 1000°C for 10 hours.

Figure 3a-b are SEM micrographs of bare and nanocrystalline cerium oxide uncoated and coated 304L stainless steel oxidized at 800°C in dry air for 168 and 442 hours, respectively. From these micrographs, it is evident that the intrinsic oxide scale developed in the unprotected 304L stainless steel is not strong enough to possess good adherence to the alloy substrate. Scale cracking and spallation are dominant under these temperature conditions. In contrast with the scale on the coated sample which is not exhibiting appreciable spallation.

Similar conclusions can be drawn from the scale morphology of 304 stainless steel samples tested at 1000°C for 10 hours in dry air. Evidently, the uncoated samples showed a coarse grained, less adherent scale and a strong tendency to spall when compared with the coated samples. Apparently, the presence of ceria as an oxygen active element modified the resultant scale rendering it more resistant to oxidation.

In addition, EDS spectra of the exposed surfaces indicated that the top scale surface of the uncoated samples are *Fe*-rich, while the bottom of the spalled zones are *Cr*-rich. This indicates that outward cation diffusion processes governs the oxidation of the bare 304 stainless steel at 1000°C in dry air. EDS spectra taken from the coated sample reveals that the top scale surface is *Cr*-rich, indicating that outward Fe cation diffusion is drastically blocked in the presence of nanocrystalline cerium oxide.

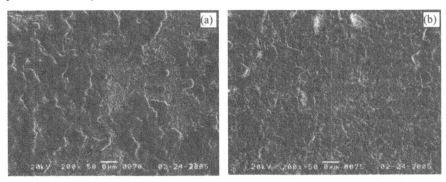

Figure 3. SEM micrographs of grade 304L stainless steel oxidized at 800°C in dry air. (a) Uncoated sample oxidized for 168 hours and (b) coated sample oxidized for 442 hours.

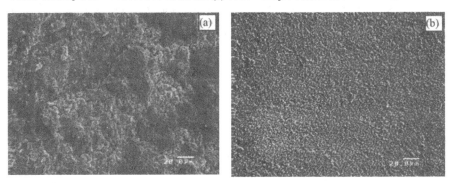

Figure 4. SEM micrographs of 304 stainless steel oxidized at 1000°C for 10 hours in dry air. (a) uncoated and (b) coated samples.

The experimental outcome of this work clearly shows that when CeO_2 nano-particles are used as coatings for oxidation protection in 304L stainless steel, there are significant improvements in the oxidation resistance exhibited by these steels. Apparently, nano-CeO_2 particle based coatings are highly effective in modifying the active oxidation mechanisms.

Among possible alternate oxidation mechanisms are: (a) CeO_2 nanoparticles peg onto grain boundaries and form complex interfaces which may act as barriers for cation and anion diffusion, (b) Once fine-grained oxide scales form, the scale growth mechanisms shift from outward cation diffusion to oxygen inward diffusion.

It is known that a minimum Cr content of approximately 20% is needed to for the formation of a protective continuous Cr_2O_3 scale on stainless steels. However, the critical minimum concentration might be significantly higher than that as the Cr supply to the alloy/oxide interface must be taken into consideration. When CeO_2 nanoparticles are placed on the steel surface they are likely to provide extra nucleation sites for Cr_2O_3 formation and, hence, complete surface coverage by Cr_2O_3 can be attained than otherwise. Thus, the critical minimum concentration of Cr needed to form a continuous protective scale on the stainless steel is reduced in the presence of CeO_{2-x} nanoparticles. Once the fine-grained oxide scales form, the scale growth mechanisms are likely to change. It is well known that oxide scale growth is dominated by cation outward diffusion as diffusivity of oxygen anions is rather low. Nevertheless , the grain boundary diffusivity of O anions is expected to be relatively high. Consequently, the oxidation mechanism might shift to O inward diffusion in the presence of CeO_2 as this is consistent with the EDS analyses.

Microscopic Simulation

Cerium-oxide-based materials possess a fluorite structure (see Fig. 5), with cerium and oxygen atoms occupying cubic and tetrahedral sites, respectively. Assuming that the protective coatings are based on the fluorite structure, computations are made by considering a double cell in which a single vacancy is present at the center of the cell (Fig. 5). In this case, the generalized gradient approximation (GGA) from Perdew and Wang is used for the exchange-correlation potential. Core electrons are represented by projector augmented-wave (PAW) pseudo-potentials, whereas valence electron wave-functions are expanded as a plane-wave basis set, taking a 400eV cutoff energy. The atomic coordinates are relaxed until the energy change between two ionic steps is smaller than 10^{-4}eV. The Kohn-Sham equations are solved self-consistently using a lattice parameter of 5.411Å. The effect of temperature is simulated by expanding the lattice parameter (5.411Å at 298K and 5.485Å at 1273K). The computations are implemented in a GAUSSIAN 03 package.

Alternatively, computations are carried out assuming that a chromium based coating with the corundum structure is initially developed. Corundum structure supercells with k-points grids of size (8 primitive cells, 80 atoms are used for computation purposes. Similar computational details are adopted as with the first assumption (i. e fluorite structure), but taking a 350eV cutoff energy. Figure 6 shows the corundum structure. The lattice parameters used in this case are 5.359Å at 298K and 5.406Å at 1273K.

Calculations assuming a chromia lattice alone clearly indicate that the activation energy for oxygen inward diffusion through the chromia oxide is indeed around 1.10eV at room temperature and 0.91eV at 1273K whereas for iron outward diffusion it is nearly 0.90eV and 0.78eV, respectively. This in turn suggests that iron outward diffusion dominates the development of the oxide scale on the bare substrate and it is consistent with experimental outcome.

Alternatively, in order to gain insight into the intrinsic characteristics of the applied coating, a comparison is made of the calculated activation energies for diffusion through cerium oxide alone and through chromium oxide doped with cerium. In the case it is found that the activation energy for oxygen diffusion through the cerium oxide alone is 1.12eV at room temperature and 0.96eV at 1273K, which is very close to the case of chromium oxide alone. In turn, this implies that cerium oxide alone is not able to provide better high-temperature-oxidation resistance when compared to chromium oxide alone. Moreover, the activation energy for iron diffusion is higher than for oxygen diffusion indicating that oxygen diffusion can be the dominant mechanism in the scale formation.

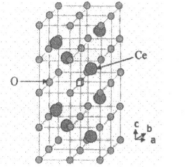

Figure 5. Schematic representation of Ce_8O_{15} tetragonal cell with an oxygen vacancy at the center.

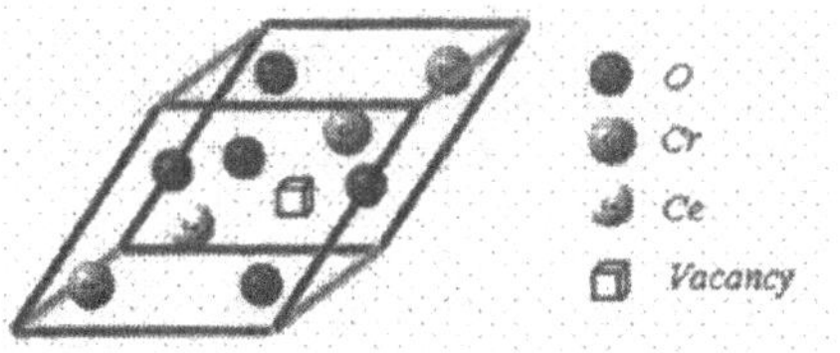

Figure 6. Schematic representation of Cr_3CeO_5 primitive cell.

Table IV. Computed Activation Energies

Temperature (K)	Activation Energy E_a (eV)					
	Ce oxide alone		Cr oxide doped with Ce		Cr oxide alone	
	O inward diffusion	Fe outward diffusion	O inward diffusion	Fe outward diffusion	O inward diffusion	Fe outward diffusion
298	1.12	1.17	1.32	1.55	1.10	0.90
1273	0.96	0.92	1.19	1.30	0.91	0.78

Table IV shows the computed activation energies from the microscopic simulations for the assumed nanoceria coatings. Notice from this table that there is a significantly large barrier to diffusion when the chromium oxide is doped with cerium, either compared with cerium oxide or chromium oxide alone. Increasing activation energies would hinder the diffusivity of oxygen and iron through the protective film, leading to a reduction in the high-temperature-oxidation rates. Furthermore, the energy barrier of iron diffusion through cerium doped chromium is higher than that of oxygen diffusion. The difference between these energy barriers leads to the conclusion that in the presence of cerium, the diffusion mechanism changes from iron outward diffusion to oxygen inward diffusion. This conclusion is supported by the experimental outcome.

CONCLUSIONS

Nanocrystalline cerium oxide particles are made by the microemulsion method with a particle size less than $5nm$. The nanocrystalline cerium oxide coatings promoted significant improvements (of 1-3 orders of magnitude) in the high-temperature-oxidation resistance of the 304 stainless steels. In particular, with a decrease in nanoparticle size, the protective effect of the nanocoatings became more dramatic.

The oxidation scales on coated samples are fine-grained, well-adherent and compact in comparison with the uncoated ones. Apparently, Cerium oxide nanoparticles provide extra nucleation sites for Cr_2O_3 formation and, consequently, complete surface coverage by Cr_2O_3 is attained more easily than in the bare material.

Activation energies for various mass transport rate limiting steps are computed using the DFT method. It is found that there is a large diffusion barrier for diffusion when the chromium oxide scale is doped with cerium. In particular, the energy barrier for iron diffusion through cerium doped chromium is higher than for oxygen diffusion. Apparently, when cerium oxide is present, the dominant diffusion mechanism changes from Fe outward diffusion to oxygen inward diffusion.

REFERENCES

1. J. Stringer, *Mater. Sci. Eng.*, A120 (1989), 129.
2. H. Buscail and J. P. Larpin, *Solid State Ion.*, 92 (1996), 243.
3. J. Shen, L. Zhou, and T. Li, *J. Mater. Sci.*, 33 (1998), 5815.
4. F. Czerwinski and W. Smeltzer, *Oxid. Met.*, 40 (1993), 503.
5. F. Czerwinski and J. Szpunar, *J. Sol-Gel Sci. and Tech.*, 9 (1997), 103.
6. S. Patil, S. C. Kuiry, S. Seal and R. Vanfleet, *J. Nanopart. Res.*, 4 (2002), 433.
7. Zhonghua Wu et al, *J. Phys.: Condensed Matter*, 13 (2001), 5269.
8. S. Patil, S. C. Kuiry and S. Seal, *Proc. R. Soc. Long.* A 460 (2004), 3569.

Mater. Res. Soc. Symp. Proc. Vol. 1243 © 2010 Materials Research Society

Hot deformation behavior of low carbon advanced high strength steel (AHSS) microalloyed with boron

I. Mejía[1], S. González-Sala[2] and J.M. Cabrera[2, 3]

[1] Instituto de Investigaciones Metalúrgicas, Universidad Michoacana de San Nicolás de Hidalgo, Edificio "U", Ciudad Universitaria, 58066–Morelia, Michoacán, México.
[2] Departament de Ciència dels Materials i Enginyeria Metal·lúrgica, ETSEIB–Universitat Politècnica de Catalunya, Av. Diagonal 647, 08028–Barcelona, Spain.
[3] CTM Centre Tecnològic, Av. de las Bases de Manresa, 1, 08240–Manresa, Spain.

ABSTRACT

This research work deals the influence of boron content on the high temperature deformation behavior of a low carbon advanced high strength steel (AHSS). For this purpose high temperature tensile and compression tests are carried out at different temperatures and constant true strain rates by using an Instron testing machine equipped with a radiant cylindrical furnace. Tensile tests are carried out at different temperatures (650, 750, 800, 900 and 1000°C) at a constant true strain rate of 0.001 s^{-1}. Uniaxial hot compression tests are also performed over a wide range of temperatures (950, 1000, 1050 and 1100°C) and constant true strain rates (10^{-3}, 10^{-2} and 10^{-1} s^{-1}). In general, experimental results of hot tensile tests show an improvement of the hot ductility of the AHSS microalloyed with boron, although poor ductility at low temperatures (650 and 750°C). The fracture surfaces of the AHSS tested at temperatures showing the higher ductility (800, 900 and 1000°C) indicate that the fracture mode is a result of ductile failure, whereas in the region of poor ductility the fracture mode is of the ductile–brittle type failure. On the other hand, experimental results of hot compression tests show that both peak stress and peak strain tend to decrease in the AHSS microalloyed with boron, which indicates that boron generates a sort of solid solution softening effect in similar a way to other interstitial alloying elements in steel. Likewise, hot flow curves of the AHSS microalloyed with boron show an acceleration of the onset of dynamic recrystallization (DRX) and a delay of the recrystallization kinetics. Results are discussed in terms of boron segregation towards austenitic grain boundaries and second phase particles precipitation during plastic deformation and cooling.

INTRODUCTION

Recent years have seen many new developments in steel technology and manufacturing processes to build vehicles of reduced mass and increased safety with steel. For this purpose Advance High Strength Steels (AHSS) have been developed. These AHSS include newer types of steels such as dual phase (DP), transformation-induced plasticity (TRIP), complex phase (CP), and martensitic (MART) steels, which are primarily multi-phase steels, which contain ferrite, martensite, bainite, and/or retained austenite in quantities sufficient to produce unique mechanical properties[1, 2]. Some types of AHSS have a higher strain hardening capacity resulting in a strength-ductility balance superior to conventional steels e.g., CP steels are characterized by high energy absorption and high residual deformation capacity. This creates an extreme grain refinement by retarded recrystallization or precipitation of microalloying elements [1]. However in this type of steels no systematic investigations have been carried out to check the beneficial effect that boron additions can promote. In boron treated steels, solute boron segregates to the austenite/ferrite grain boundaries, and increases hardenability by suppressing the ferrite transformation through strong interaction with lattice defects such as dislocations and vacancies. The delaying effect on the ferrite transformation promotes instead bainitic formation in hot rolled microalloyed steels, where the high strength is derived from accumulative effects such as fine grain size, solid solution strengthening with additional interstitial hardening, precipitation hardening by carbide and nitride particles, transformation hardening and strain hardening [3-10]. Boron has been considered as another element likely to enhance the hot ductility in a similar grade to Ti, Ca, Zr and Y [11-13]. At relatively low temperatures, where austenite and ferrite phases coexist, such beneficial effect of boron has been suggested to be due to the precipitation of $Fe_{23}(B,C)_6$ in the matrix [14]. In hot working processes, the effect of boron on the hardenability of austenitic steel is pronounced [4-10, 15], particularly its interaction with grain boundaries, which is relevant to hot working and recrystallization of austenite. Segregated boron atoms on austenite grain boundaries will change their interfacial energy and this in turn exerts some effect on the softening behavior of austenite during hot working [16]. At present, there are just a few studies strictly focused on the boron effect on the hot ductility and hot flow behavior of AHSS steels [14, 17-22]. Therefore, the aim of this paper is to study the effect of boron content on the hot ductility of a low carbon advanced high strength steel and their flow behavior by analysis of its influence on the high temperature flow curves.

EXPERIMENTAL METHOD

Two low carbon AHSS are prepared in the Foundry Laboratory of the Metallurgical Research Institute of UMSNH by using high purity raw materials in a 25 kg capacity induction furnace and cast into 70x70 mm cross section ingots. One of the steels is microalloyed with boron. The chemical composition of both steels (see table 1) seems to be homogeneous enough to undertake an accurate study on the effect of boron content on this steel.

Table 1. Chemical composition of the low carbon advanced high strength steels (wt %).

AHSS*	C	Mn	Si	Cr	Ni	Cu	V	B
B0	0.032	0.367	0.260	1.144	1.930	0.388	0.195	0.0000
B5	0.035	0.372	0.239	1.130	1.910	0.387	0.193	0.0117
*As hot rolled+quenched condition: σ_{max} > 900 MPa, bainite-martensite CP-AHSS steel.								

To evaluate the high temperature ductility, cylindrical hot tensile specimens of 6 mm in diameter and 30 mm in gauge length are machined from the as cast condition steel ingots. Isothermal hot tensile tests are performed at different temperatures and at a constant true strain rate (10^{-3} s^{-1}) using an Instron testing machine, equipped with a radiant cylindrical furnace. All the tests are performed in an argon atmosphere in order to protect the molybdenum-based tools from oxidation as well as to prevent oxidation of the samples. First, the specimens are heated to 1100°C and held for 900 s in order to homogenize the microstructure, and to obtain a similar initial grain size. Then specimens are cooled down to the test temperatures (650, 750, 800, 900 and 1000°C) and held again for 300 s. Immediately after rupture, the broken specimens are quenched by an argon stream and finally the fractures surfaces are examined by scanning electron microscopy (SEM). Hot ductility is quantified by the % of reduction in area (RA) at fracture. To estimate the hot flow behavior, cylindrical specimens of 7 mm in diameter and 11 mm in length are machined from the as cast condition steels ingots. Isothermal hot compression tests are then performed at different temperatures and strain rates in the same machine above mentioned. All tests are also performed under an argon atmosphere and preheated according to the same procedure of tensile tests. Then specimens are cooled down to the test temperature (950, 1000, 1050 and 1100°C) and held again for 300 s. Specimens are strained until $\varepsilon = 0.8$ at different constant true strain rates (10^{-3}, 10^{-2} and 10^{-1} s^{-1}). This wide range of conditions provided a good description of the hot flow behaviour of steels. The soaking conditions promoted an initial austenitic grain size ranging from 30.8 ± 21.7 μm in B0 steel (0 ppm B) to 40.1 ± 36.7 μm in B5 steel (117 ppm B). Therefore, in this work the initial austenitic grain size has been considered similar for both steels and it will be assumed that no influence of this microstructural parameter on the hot flow behavior must be expected.

RESULTS AND DISCUSSION

Hot ductility evaluation

The effect of deformation temperature on the tensile behavior of the low carbon AHSS is shown in figure 1. As expected, the strength decreases as temperature increases for both steels. It must be noticed that there are fluctuations in the flow curves at 1000°C, particularly in the low carbon AHSS microalloyed with boron (B5), which are evidence of single peak DRX [23], a fact later confirmed by the hot compression tests. However, no oscillations are observed at lower temperatures, i.e. at 650, 750 and 800°C. In terms of elongation, both flow curves at 650°C display the lowest values. A more convenient way of measuring the hot ductility is through the reduction of area (RA) in fractured samples. Accordingly, RA dependence on the deformation temperature for both steels is shown in figure 2, where one can notice that RA decreases with decreasing temperature, and the typical recovery of hot ductility at lower temperatures (650 and 750°C) associated with absence of recrystallization does not appear. In this case the low carbon AHSS microalloyed with boron (B5) shows the higher hot ductility in all range of tested temperatures. According to Mintz et al. [19], the temperature range in which the RA is lower than or equal to 60% is a crack sensitive range for continuous casting, and it is called the hot brittle range. With this rule, and in the present steels, when boron is added, the hot brittle range tends to disappear leading to a further improvement in hot ductility. The high ductility observed at 900 and 1000°C is associated to the thermal activation of two simultaneous phenomena, namely grain boundary sliding and DRX [24]. These two mechanism move grain boundaries away from microcracks, keeping them isolated and preventing in turn their coalescence. Therefore, the higher hot ductility in the low carbon AHSS microalloyed with boron is related to the stress accommodation by extensive plastic flow rather than cracking at grain boundaries [24]. Accordingly, Lagerquist and Langenborg [25] show that boron additions to an alloy can reduce grain sliding and improve creep ductility.

It has been reported that the segregation of impurities to grain boundaries plays an important role in the loss of hot ductility in addition to the thin pro-eutectoid ferrite layer formation and the precipitation of carbides or nitrides of V, Ti, Nb, Al or B at austenite boundaries [26-34].

31

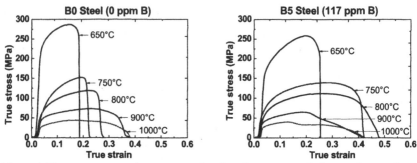

Figure 1. True stress-true strain curves as a function of temperature during tensile test of examined low carbon AHSS.

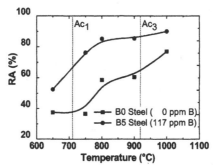

Figure 2. Hot ductility curves for the low carbon AHSS (RA reduction in area).

In the case of Cu containing steels, the problem of hot cracking is known as hot shortness and is caused by the formation of a liquid film of Cu at the austenite grain boundaries, which penetrates along grain boundaries and causes cracking during hot processing under oxidizing conditions [35]. Mintz *et al.* [36] have shown that the poor hot ductility is due to the precipitation of fine CuS particles and that the addition of Ni by increasing the solubility of Cu in austenite reduces the driving force for their precipitation. In this way, the variables which have a major influence on the poor hot ductility are precipitation and inclusions; the finer the precipitation, the worse is the ductility, also boundary precipitation is particularly deleterious [26]. In the case of Nb and V containing steels, a large part of the precipitation comes out dynamically during the straightening operation and so can be very detrimental to hot ductility [26].

Fractography

The fracture surfaces of the low carbon AHSS tested at 650, 800, and 1000°C are shown in figures 3, 4, and 5, respectively. At higher temperatures (1000°C) fracture seems to be ductile with small and large dimples and without apparent grain boundary facets. The fracture surface appearance coincides with the higher hot ductility reported by other authors [14, 24]. Voids and small cracks can be formed at the ferrite-austenite interface, leading to low ductility by intergranular failure, as seen from fracture at 650°C. In this latter case, the fracture is primarily the result of a ductile failure, with some brittle characteristics. There are also some voids and cavities formed at grain boundaries and precipitates or inclusions are found within the voids. The intergranular separation is more prominent at lower temperatures. The large improvement in the hot ductility of the examined low carbon AHSS microalloyed with boron can be due mainly to the increase in grain boundary cohesion. This arises from grain boundary segregation of boron [37-42], which can occupy proper nucleation sites preferentially and lower austenite grain boundary energy, suppressing the decomposition of austenite and increasing grain boundary cohesion [43].

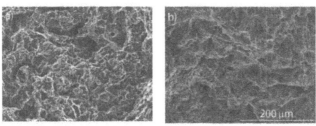

Figure 3. SEM fractographies of specimens tensile-tested at 650°C. (a) B0 steel, (b) B5 steel.

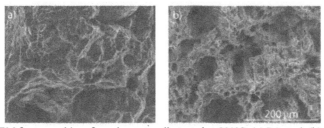

Figure 4. SEM fractographies of specimens tensile-tested at 800°C. (a) B0 steel, (b) B5 steel.

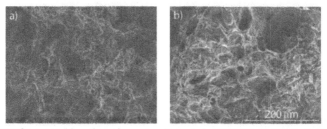

Figure 5. SEM fractographies of specimens tensile-tested at 1000°C. (a) B0 steel, (b) B5 steel.
Hot flow behavior

Figure 6 shows the flow curves of the hot compression tests of the low carbon AHSS.

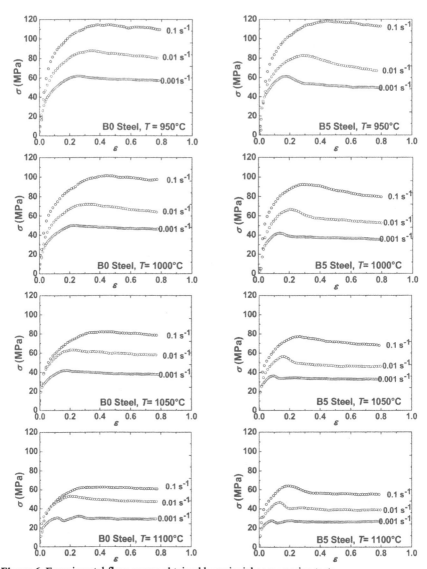

Figure 6. Experimental flow curves obtained by uniaxial compression tests.

These flow curves exhibit the expected behavior when DRX occurs [23,44]. Both steels show a similar hot flow behavior, with the classic dependence of peak stress (σ_p) and peak strain (ε_p) on temperature and strain rate, i.e. σ_p and ε_p increase as strain rate increases and temperature decreases. In all cases the steel microalloyed with boron (B5) shows lower peak stress values than the steel without boron (B0), as shown in figure 7. Similar remarks can be done about the steady state stresses. It is also observed that the softening beyond the peak stress value is more pronounced in the B5 steel.

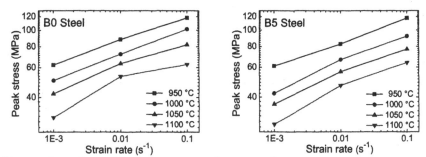

Figure 7. Dependence of the peak stress on strain rate at each temperature.

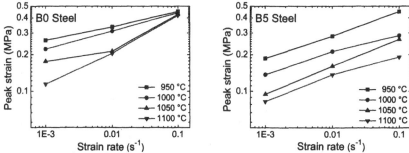

Figure 8. Dependence of the peak strain on strain rate at each temperature.

In this study it is evident that boron additions produce softening and also a delay of the DRX during the hot deformation of the steel (see figure 6). This delaying effect of the DRX has already been reported by other authors in previous studies [21, 45-48]. The authors consider that this feature indicates that boron additions could promote a solid solution softening effect additional to the DRX itself, in a similar way to the role played by other interstitial alloying elements such as carbon in steels [49-51]. This solid solution softening effect can be associated with diffusion and segregation of the boron atoms towards the austenitic grain boundaries where they occupy the vacancies generated by the applied deformation and modify and reinforce the cohesion of austenitic grain boundaries; this fact would allow an easier plastic flow in the austenitic lattice [24, 34, 45]. The softening effect shown by the present boron microalloyed steel at high temperatures is opposite to that shown at low temperatures, where increasing the boron content increases the hardness of steel [8].

35

To reach the onset of DRX it is imperative to provide a critical strain to begin the recrystallization process. After the peak stress is attained, an additional softening is observed up to a point in which the steel achieves a steady state stress. This state can be obtained in two different ways. At relatively low strain rates and high temperatures, the flow curves show a cyclic behavior [23] (repeated oscillation in the flow curve, as noticed in figure 6 for the B5 steel tested at 0.001 s^{-1} and 1100°C), while at high strain rates and low temperatures, only a single peak stress is observed (as displayed in figure 6 for the B5 steel tested at 0.1 s^{-1} and 950°C). As illustrated in figure 6, DRX preferably appears at strain rates of 0.01 and 0.001 s^{-1} and at relatively high temperatures (1000, 1050 and 1100°C). Under these conditions, Schulson et al. [52] have reported that boron improves the mobility of grain boundary dislocations during hot deformation which in turns facilitate DRX. This would explain partially the present acceleration of the onset of DRX (diminution of ε_p, as can be seen in figure 8) at increasing boron content. As mentioned before, boron atoms can also segregate towards austenitic grain boundaries and occupy the sites of vacancies generated by deformation avoiding the formation and propagation of microcraks [50]. This increases the cohesion among austenitic grains and facilitates the plastic flow and deformation by slip accommodation [24]. On the other hand, an additional important effect is detected in these flow curves. A delay of DRX kinetics takes place at increasing boron content. This can be noted in the time spent in attaining the steady state stress once DRX has started, i.e. on the shape of the flow curve after the onset of DRX. The delay of the DRX kinetics is usually associated to a solute drag effect [23].

CONCLUSIONS

Boron additions to the present low carbon advanced high strength steel are beneficial to the hot ductility. The improved hot ductility of the boron microalloyed steel is associated to the enhanced grain boundary cohesion and an easier flow in the austenite lattice. On the other hand, the typical recovery of hot ductility at lower temperatures associated with absence of recrystallization does not appear in the present steels. This is associated to precipitation and inclusions, particularly CuS particles and vanadium carbides/nitrides. The fracture surface of the boron microalloyed steel tested at temperatures showing the higher hot ductility indicate that fracture mode is a result of ductile failure, whereas in the region of poor hot ductility the fracture mode is ductile-brittle failure.

Hot flow curves of the present low carbon advanced high strength steels show that both peak stress and peak strain tend to decrease when boron is added, which indicates that boron additions generate a solid solution softening effect in a similar way to the hot flow behavior of other interstitial alloying elements like carbon. Likewise, hot flow curves of boron microalloyed steel show an acceleration of the onset of DRX and a delay of the recrystallization kinetics.

ACKNOWLEDGMENTS

I. Mejía acknowledges the National Council of Science and Technology from México (CONACyT) for the financial support for this project. All the authors, also acknowledge Departament de Ciència dels Materials i Enginyeria Metal·lúrgica of the Universitat Politécnica de Catalunya (Spain), for the support and technical assistance in this research work, especially Montse Marsal for her support in the SEM facilities. Funding is obtained through project

CICYT-DPI 2005-09324-C02-01 (Spain) and Coordinación de la Investigación Científica of the UMSNH (México).

REFERENCES

1. Committee on Automotive Applications, International Iron & Steel Institute, *Advanced High Strength Steel Application Guidelines*, 1-9 (2006).
2. L. Duensing, *Modern Metals* **62**, 92 (2006).
3. R.D.K. Misra, G.C. Weatherly, J.E. Hartmann and A.J. Boucek, *Mater Sci. Technol.* **17**, 1119 (2001).
4. K.E. Thelning, *Steel and its Heat Treatment*, (Butterworths, London, 1984).
5. W.C. Leslie, *The Physical Metallurgy of Steels*, (McGraw-Hill, USA, 1981).
6. B.M. Kapadia, in *Hardenability Concepts with Application to Steel*, edited by D.V. Doane and J.S. Kirkaldy, (The Metallurgical Society of the AIME, Warrendale, PA., USA, 1978) pp. 448-480.
7. D.A. Mortimer and M.G. Nicholas, *Met. Sci.* **10**, 326 (1976).
8. J.E. Morral and J.B. Cameron, *Met. Trans. A* **8**, 1817 (1977).
9. Ph. Maitrepierre, D. Thivellier, J. Roves-Vernis, D. Rousseau and R. Tricot, in *Hardenability Concepts with Application to Steel*, edited by D.V. Doane and J.S. Kirkaldy, (The Metallurgical Society of the AIME, Warrendale, PA, USA, 1978) pp. 421-447.
10. R.C. Sharma and G.R. Purdy, *Met. Trans.* **5**, 939 (1974).
11. O. Comineli, R. Abushosha and B. Mintz, *Mater Sci. Technol.* **15**, 1058 (1999).
12. B. Mintz, Z. Mohamed and R. Abushosha, *Mater Sci. Technol.* **5**, 682 (1989).
13. K. Suzuki, S. Miyagawa, Y. Saito and K. Shiotani, *ISIJ Int.* **35**, (34 1995).
14. S.K. Kim, J.S. Kim and N.J. Kim, *Metall. Mater. Trans. A* **33**, 701 (2002).
15. J.E. Morral and J.B. Cameron, in *Boron Hardenability Mechanisms, Boron in Steels*, edited by S.K. Banerji and J.E. Morral, (Proc. of the Metallurgical Society of AIME, Milwaukee, Wisconsin, USA, 1979) pp. 19.
16. E.D. Hondros and M.P. Seah, *Int. Met. Rev.* **22**, 262 (1977).
17. H.-W. Luo, P. Zhao, Y. Zhang and Z.-J. Dang, *Mater. Sci. Technol.* **17**, 843 (2001).
18. J.N. Tarboton, L.M. Matthews, A. Sutcliffe, C.M.P. Frost and J.P. Wessels, *Mater. Sci. Forum* **318-3**, 777 (1999).
19. B. Mintz and R. Abushosha, *Mater. Sci. Technol.* **8**, 171 (1992).
20. E. Lopéz-Chipres, I. Mejía, C. Maldonado, A. Bedolla-Jacuinde and J.M. Cabrera, *Mater. Sci. Eng. A* **460-461**, 464 (2007).
21. E. López-Chipres, I. Mejía, C. Maldonado, A. Bedolla-Jacuinde, M. El-Wahabi and J.M. Cabrera, *Mater. Sci. Eng. A* **480**, 49 (2008).
22. I. Mejía, E. López-Chipres, C. Maldonado, A. Bedolla-Jacuinde and J.M. Cabrera, *Int. J. Mat. Res. (formerly Z. Metallkd.)* **99**, 12 (2008).
23. T. Sakai and J.J. Jonas, *Acta Metall.* **32**, 189 (1984).
24. F. Zarandi and S.Yue, *ISIJ Int.* **46**, 591 (2006).
25. M. Lagerquist and R. Langenborg, *Scand. J. Metall.* **1**, 81 (1972).
26. B. Mintz, S. Yue and J.J Jonas, *Int. Mater. Rev.* **36**, 187 (1991).
27. W.T. Nachtrab and Y.T. Chou, *J. Mater. Sci.* **19**, 2136 (1984).
28. Y. Maehara, K. Yasumoto, H. Tomono, T. Nagamichi and Y. Ohmori, *Mater. Sci. Technol.* **6**, 793 (1990).

29. W.T. Nachtrab and Y.T. Chou, *Metall. Trans. A* **17**, 1995 (1986).

30. W.T. Nachtrab and Y.T. Chou, *Metall. Trans. A* **19**, 1305 (1988).

31. R. Abushosha, R. Vipond and B. Mintz, *Mater. Sci. Technol.* **7**, 1101 (1991).

32. H. Matsuoka, K. Osawa, M. Ono and M. Ohmura, *ISIJ Int.* **37**, 255 (1997).

33. C. Nagasaki and J. Kihara, *ISIJ Int.* **37**, 523 (1997).

34. S.-H. Song, A.-M. Guo, D.-D. Shen, Z.-X. Yuan, J. Jiu and T. -D. Xu, *Mater. Sci. Eng. A* **360**, 96 (2003).

35. B. Mintz, *ISIJ Int.* **39**, 833 (1999).

36. B. Mintz, R. Abushosha and D. N. Crowther, *Mat. Sci. Technol.* **11**, 474 (1995).

37. T.-D. Xu, S.-H. Song, Z.-X. Yuan and Z.-S. Yu, *J. Mater. Sci.* **25**, 1739 (1990).

38. S.-H. Song, T.-D. Xu, Z.-X. Yuan and Z.-S. Yu, *Acta Metall. Mater.* **39**, 909 (1991).

39. T.-D. Xu, S.-H. Song, H.-Z Shi. Z.-X Yuan and W. Gust, *Acta Metall. Mater.* **39**, 3119 (1991).

40. X.L. He, Y.Y. Chu and J.J. Jonas, *Acta Metall. Mater.* **37**, 147 (1989).

41. B. Cao, X.-W. Wang, H.-Y Cui and X.L. He, *J. Univ. Sci. Technol.* **B9**, 347 (2002).

42. Z.L. Zhang, Q.-Y. Lin and Z.-S. Yu, *Mater. Sci. Technol.* **16**, 305 (2000).

43. M.P. Seah, *Acta Metall. Mater.* **28**, 955 (1980).

44. F.J. Humphreys and M. Hatherly, *Recrystallization and Related Annealing Phenomena*, (Pergamon Press, Oxford, 1995).

45. X.L. He, M. Djahazi, J.J. Jonas and J. Jackman, *Acta Metall. Mater.* **39**, 2295 (1991).

46. M. Djahazi, X.L. He, J.J. Jonas and L. Collins, in *Recrystallization '90*, edited by T. Chandra, (TMS-AIME, 1990) pp. 681.

47. M. Djahazi, X.L. He and J.J. Jonas, in *Proc. Int. Conf. on Phys. Metall. of Thermomechanical Processing of Steels and Other Metals*, edited by I. Tamura, (Thermec-88 (1), Tokyo, Japan, 1988) pp. 246.

48. M. Djahazi, X.L. He, J.J. Jonas and G.E. Ruddle, in *Proceedings of the International Symposium on Processing, Microstructure and Properties of HSLA Steels*, edited by A.J. Deardo, (TMS-AIME, Warrendale, PA, 1988) pp. 69.

49. T. Sakai, Z. Xu and G.R. Zhang, *Tetsu-to-Hagané* **80**, 557 (1994).

50. Z. Xu, G.R. Zhang and T. Sakai, *ISIJ Int.* **35**, 210 (1995).

51. F. Escobar, J.M. Cabrera and J.M. Prado, *Mater. Sci. Technol.* **19**, 1137 (2003).

52. E.M. Schulson, T.P. Weihs, D.V. Viens and I. Baker, *Acta Metall. Mater.* **33**, 1587 (1985).

Mater. Res. Soc. Symp. Proc. Vol. 1243 © 2010 Materials Research Society

Synthesis of Fullerene by Spark Plasma Sintering and Thermomechanical Transformation of Fullerene Into Diamond on Fe-C Composites

Francisco C. Robles-Hernández[1†] and H. A. Calderon[2‡]

[1] University of Houston, Engineering Technology, Houston, TX, USA 77204
[2] Departamento Ciencia de Materiales, ESFM-IPN, Mexico DF

ABSTRACT

In this work, results are presented regarding the characterization of nanostructured Fe matrix composites reinforced with fullerene. The fullerene is a mix of 15 wt.%C_{60}, 5 wt.%C_{70} and 80 wt.% soot that is the product of the primary synthesis of C_{60}. The composite has been produced by means of mechanical alloying and sintered by Spark Plasma Sintering (SPS). The characterization methods include XRD, SEM and TEM. The C_{60} and C_{70} withstand mechanical alloying, SPS, and thermomechanical processing and act as a control agent during mechanical alloying. The results show that the mechanically alloyed and SPS product is a nanostructured composite. A larger amount of C_{60} is found in the sintered composite than in the original fullerene mix, which is attributed to an in-situ synthesis of C_{60} during the SPS process. The synthesis of C_{60} is presumably assisted by the catalytic nature of Fe and the electric field generated during the SPS process. In order to study the effect of high temperature, high strain, high heating and cooling rates on C_{60}, the composite is subjected to a thermomechanical processing; demonstrating that some of the C_{60} resists the above described environment and some of it partially transforms into diamond.

INTRODUCTION

The synthesis of fullerene and carbon nanotubes has been successfully reported by different methods, the resulting structures include buckyballs, buckytubes, onions, giant fullerene structures, concentric structures, etc. [1-5]. The most common structures are the buckyballs (C_{60} and C_{70}) formed by atomic arrangements of carbon in networks of pentagons, hexagons [2], and heptagons [6]. Fullerenes are usually synthesized and purified by the method proposed by Krätschmer [7]. As of today, the main focus has been given to characterization and applications; e.g. fullerenes have been used as reinforcements for structural materials [8-10].

Fullerenes are susceptible to transform by mechanical and heat treatment means [11-13] and can result in significant technological as well as scientific development. Previous research indicate that C_{60} and C_{70} can resist mechanical milling (in a ball mill) for up to 1000 h without amorphization; in contrast the milled C_{60} form polymer-like networks or molecules such as C_{120} [11]. Metallic and ceramic matrix composites can be successfully produced by mechanical alloying and mechanical milling, including ductile metals [8-10, 14-17]. The alloyed powders are usually highly homogeneous and nanostructured at low temperatures. Spark Plasma Sintering (SPS) is the ideal method to preserve the nanostructured nature of mechanically alloyed powders and allows an almost complete densification of the materials [10].

Corresponding authors, Emails: † fcrobles@uh.edu and ‡ hcalder@esfm.ipn.mx

Some of the success for mass production of nanotubes is attributed to the catalytic effect of transition metals [12, 13, 18-21]. The first report indicating the presence of multiwalled carbon nanostructures appeared during the 1950's. In such report, it is found that the carbon soot from a brick's sintering chimney contains unique carbon nanoparticles with rod-like appearance. Such rod-like particles had been formed along the brick's cavities in the vicinity of iron rich regions that, presumably, catalyze its formation [1]. Production of MoS_2 nanorods is possible due to the catalytic effect of Mo, with the novelty that these nanorods are synthesized in a conventional resistance furnace [12,13]. This means that no electric arc is present during the synthesis of the MoS_2 nanorods. Carbon nanotubes can be end opened by chemical means [24] and can be filled with elements or compounds [22,23], used as nanoextruders [25], pressure cells or vessels (carbon "onions") [5]. While as synthesized (closed) nanotubes have been used as speakers [26], for composites [27,28], biosensors [29], and other applications.

In the present work mechanical alloying and SPS are combined to produce a nanostructured composite of the Fe-fullerene system. The Fe-fullerene composite is subjected to a thermo-mechanical processing (high temperature, high strain-high stress) and the characterization indicates that the fullerene mix transforms into various phases including diamond. Characterization is conducted by means of Scanning Electron Microscopy (SEM), X-Ray Diffraction (XRD), Transmission Electron Microscopy (TEM), and micro-hardness. The results are reported and discussed accordingly.

EXPERIMENTAL PROCEEDURE AND MATERIALS

Powders of iron (Fe; 99.9% purity and a particle size <100μm), and fullerene mix (15.5 at.% C_{60} + 4.5 at.% C_{70} + 80 at.% C_{Soot}) are used as starting materials. The nominal composition of the composite is Fe-15.7 at.% C. In the present paper the above described mix containing C_{60} + C_{70} + soot is referred as fullerene mix and the term "original sample" is used to identify the mixed powders before mechanical alloying. Mechanical alloying is carried out using a horizontal ball mill for various times between 0 and 100 h; however, due to the relevance to this manuscript only the results for 100 h of mechanical alloying are presented. For mechanical alloying a ball to powder ratio of 100:1 has been used. No control agent is added to the ball mill. The manipulation of the powders is always conducted under controlled atmosphere (Ar) environment to prevent any degradation of the powders or potential reactions with either nitrogen or oxygen before or during the milling process. After the milling process, the mills are opened slowly (in a period of 1-2 days) to prevent potential ignition of the powders.

The alloyed powders are sintered using a Doctor Sinter 1080 (SPS) apparatus under the following conditions: 773 K, 100 MPa for 600 s, with a heating rate of 3.3 K/s. The sintered products are 13 mm in diameter and 5 mm in thickness. Following the SPS process the Fe-fullerene composite is thermomechanically processed using a conventional steel rolling mill. For the thermomechanical process, samples of 1.5 x 1.5 x 10 mm are used. These samples are extracted from the sintered product by means of electro discharge. In order to facilitate the manipulation of the samples, the specimens are sandwiched within two layers of stainless steel sheet (0.5 mm in thickness). The "sandwich" is heated in a resistance furnace in open atmosphere conditions for 10 minutes at a temperature of 1273 K. The temperature inside the furnace is monitored with an Omega® "K" type thermocouple. Therefore, it is expected that before the rolling process the temperature may drop between 50 and 100 K. Each rolling is done in a single pass to reach area reductions of 21%, 43% and 70% followed by water quenching.

The X-ray diffraction (XRD) is conducted on a SIEMENS D5000 apparatus using a Cu tube with a K_α wavelength of 0.15406 nm. The SEM observations are carried out on a JEOL JSFM35CF microscope operated at 20 kV. The SEM samples in powder form are prepared by dispersing the powders onto a graphite tape. TEM is conducted on a JEOL JEM200FXII operated at 200 kV. For TEM, the samples are prepared by dispersing the powders on ethanol to form a dilute suspension from which an aliquot is taken and is deposited on a Cu-graphite 300 mesh. The as sintered and as thermomechanical products are surface polished for SEM, and electropolished on a double jet polisher for TEM. Electropolishing is conducted at 230 K using an electrolyte solution of 25 vol.% HNO_3 and 75 vol.% CH_3OH.

RESULTS AND DISCUSSION

In Figure 1, XRD results of the original sample are presented, together with those for mechanically alloyed Fe-fullerene powders after 100 h of milling. The typical reflections for pure iron can be easily noticed but the reflections of fullerene are not evident. The extinction of the fullerene reflection(s) can be attributed to a combined effect between the small amount of C_{60} + C_{70} in the fullerene mix and a potential fluorescence or absorption effect of the characteristic C_{60} + C_{70} radiation by iron. On the other hand, a reduction in the intensity for all the XRD reflections, but the (110) is observed in the milled powders. This can be the result of plastic deformation taking place during mechanical alloying that forces the iron powders to elongate in preferential directions. This results in the development of a sample with crystalline anisotropy or texture.

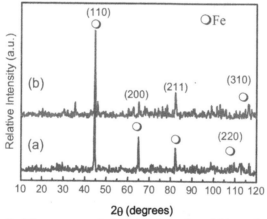

Figure 1. XRD of the Fe-fullerene composite (a) original sample and (b) sample milled for 100 h in in a horizontal mill.

SEM images are shown in Figure 2, they correspond to the original powders (iron and fullerene mix) as well as the mechanically alloyed ones. From these images, it can be seen that by using ball milling, the iron and fullerene powders form highly homogeneous and relatively smaller particles. It is also evident that the newly formed particles agglomerate and form distinctive groups of particles. They are the basis of the nanostructured composite. The highly irregular morphology of the Fe-fullerene particle groups is attributed to the control agent effect of the fullerene mix that prevents excessive agglomeration and texture formation.

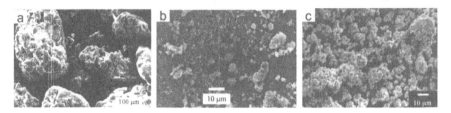

Figure 2. SEM images of powders in use, (a) iron, (b) fullerene mix (C_{60} + C_{70} + soot) and (c) Fe-fullerene composite mechanically alloyed for 100 h in a horizontal mill.

Figure 3 shows a representative TEM dark field image of the as-milled Fe-fullerene composite. The following facts can be inferred from such a figure: formation of a nanostructured Fe-fullerene composite, high homogeneity of the powders and the presence of iron and C_{60}. The particles observed in Figure 2c are agglomerations of nanostructured crystals. C_{60} in the presence of iron powders stands mechanical ball milling for 100 h, and no chemical reaction among the constituents is identified. The identification of C_{70} is challenging since its absolute amount in the fullerene mix is lower than 5%. In addition, soot can only be identified, by TEM and XRD, if large amounts of graphene layers are aligned perpendicularly to the incident beam. The table in Figure 3 shows the identification of the respective reflections of the TEM diffraction pattern (insert in Fig. 3).

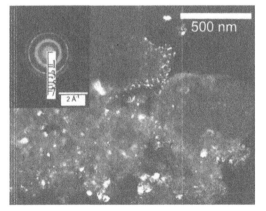

Iron Reflections

# ID	(hkl)	d(Å)
4	(110)	2.01
6	(200)	1.49
8	(211)	1.15
9	(220)	1.00

Fullerene Reflections

# ID	(hkl)	d(Å)
1	(422)	2.98 (g)
2	(440)	2.59
3	(533)	2.17
5	(820)	1.75
6	(751)	1.61
7	(931)	1.49

Figure 3. TEM dark field and electron diffraction pattern with the corresponding identification of the reflections for iron and fullerene in the 100 h mechanically ball milled powders. (g) indicates the reflection used to generate the dark field.

Figure 4 shows the XRD of the SPS and the thermomechanically processed Fe-fullerene composite. It is evident that the effect of SPS sponsors the synthesis of C_{60} fullerene. A clear and strong peak for fullerene can be found in the profile (a) of Fig. 4. The successful synthesis of C_{60} is attributed to the catalytic characteristics of Fe, one of the transition metals normally used to synthesize various forms of fullerenes and other forms of carbon nanotubes [12,13,18-21]. The other mechanism that may be playing an important role in the synthesis of C_{60} is the quasi-wrapping effect to form the so called "giant onions" that have been discovered by Ugarte [5]. The giant onions are formed when carbon nanoparticles (e.g. graphenes) are exposed to an electric field. It is likely that the synthesis of C_{60} is the result of the same quasi-wrapping mechanism of the graphenes present in the soot. Furthermore, there is a possibility that other types of fullerenes (e.g C_{20}, C_{70}, etc) are also synthesized, but in smaller amounts and this prevents their identification.

The controversy regarding the presence of plasma on the SPS process [30,31] is known to the authors, and the above mentioned quasi-wrapping mechanism is rather assisted by the electric field formed by the SPS process. This synthesis is of great importance since the final amount of C_{60} is significantly larger than that in the original fullerene mix (compare Figure 4 and Figure 1). In the present work, all the C_{60} is synthesized during the SPS process that takes approximately 10 min. In contrast, using conventional methods it is possible to synthesize only a fraction of a gram of fullerenes or other forms of carbon nanoparticles in several hours. Therefore, this finding is considered outstanding since this may facilitate the mass production of fullerene (e.g. C_{60}) or other form of carbon nanoparticles (e.g. SWCNT, MWCNT, etc).

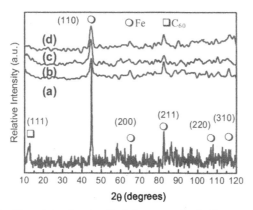

Figure 4. XRD of the (a) as sintered and as (b-d) thermomechanically processed Fe-fullerene samples using (b) 21%, (c) 43% and (d) 70% of deformation at 1273 K.

The remaining profiles (b-d) in Fig 4 correspond to the thermomechanically processed Fe-fullerene composites. Progressively higher deformations have been applied to thin sheets of sintered composite by rolling. Figure 4 (b-d) shows that the intensity of the reflections for both iron and C_{60} decrease very considerably. However, this reduction in intensity is not exactly linear with the applied deformation during the rolling process. Although the deformation is enough to affect the intensity of the reflections of iron, the presence of fullerene is in all cases still evident (see Figure 4b-d at low angles). This suggests that C_{60} stands, not only high impacts (mechanical alloying), but also high temperature-high strain environments. The background noise in the XRD of the sintered and thermomechanically processed composites prevents the identification of other phases such as Fe_3C due to its lower absolute volume fraction. Nonetheless, the presence of Fe_3C is more evident in the following TEM results (Figure 5). On the other hand, the hardness of the SPS composite increases from approximately 100 μHV (iron matrix) to 722 μHV i.e., a difference of more than 700 %. This increase in hardness is attributed to the presence of fullerenes and the formation of carbides and other hard phases.

Figure 5 shows two TEM dark field images and their respective diffraction patters for the as sintered Fe-fullerene composite. The analysis of the diffraction pattern confirms that some of the phases identified in the composite are iron and C_{60}. The diffraction patterns presented in Figure 5 indicate that the respective lattice parameters for iron and fullerene are 0.28 Å and 1.38 Å. There is a 1 % and a 4 % difference with respect to the theoretical lattice parameter values for Fe and C_{60} [2, 9, 33,], respectively, such difference in the context of phase identification by TEM techniques, is considered negligible.

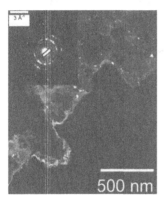
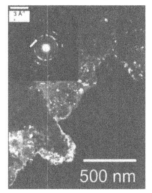

Iron Reflections

Position	(hkl)	d(Å)
4	(110)	2.01 (g)
5	(200)	1.460
6	(211)	1.147
7	(220)	1.085
8	(310)	0.873

Fullerene Reflections

Position	(hkl)	d(Å)
1	(220)	5.02
2	(420)	3.09 (g)
3	(440)	2.591
4	(444)	2.008
5	(931)	1.460

Figure 5. Dark field images and electron diffraction patters for the as sintered (SPS) composite showing the (a) reinforcement (C_{60}) and the (b) matrix (iron). Arrows indicate the reflection (g) used to generate the respective dark fields. The adjacent table shows the respective reflections for the identified phases in the composite.

Figure 6 shows additional TEM dark field images and their respective diffraction patterns for carbides and iron present in the as sintered Fe-fullerene composite. The electron diffraction pattern shows the presence of iron (matrix), C_{60} (reinforcement), and iron carbide (Fe_3C). The formation of the Fe_3C is assisted during the SPS process. The lattice parameters of the identified phases in the electron diffraction pattern shown in Figure 6 are 0.28 Å (Fe), 1.42 Å (C_{60}), and for Fe_3C: $a = 5.11$ Å, $b = 6.72$ Å, and $c = 4.55$ Å. The difference in lattice parameters with respect to other reported values is in all cases less than 1 % [2, 9, 33, 34]. This allows concluding that after SPS all the carbon present in the original fullerene mix has been transformed into C_{60} or reacted

44

to form Fe_3C. However, Fe_3C is not found in XRD (Figure 4) which is attributed to the limited amount formed during SPS together with the relatively high noise in the XRD background.

Figure 7 shows TEM images of the sintered products after thermomechanical processing at 21 % and 43 % deformation. Figure 7a-b present dark fields showing C_{60} and their respective diffraction pattern. Figure 7c shows a C_{60} particle with a tetragonal crystalline structure having the following lattice parameters $a = 0.68$ nm and $c = 1.11$ nm. Such particle has an approximate

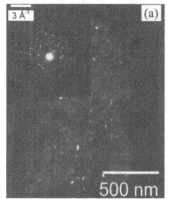
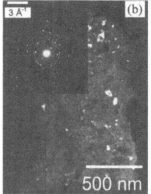

Iron Reflections		
Position	(hkl)	d(Å)
5	(110)	2.01 (g)
7	(200)	1.49
8	(211)	1.15
9	(220)	1.09
10	(310)	0.87
11	(222)	0.80

Fullerene Reflections		
Posición	(hkl)	d(Å)
1	(220)	5.02
3	(422)	2.87

Iron Carbide (Fe₃C) Reflections		
Posición	(hkl)	d(Å)
2	(002)	3.36 (g)
4	(004)	2.59
6	(114)	1.60

Figure 6. Dark field images and electron diffraction patters for the as sintered composite showing (a) Fe_3C (cementite) and (b) iron matrix. Arrows indicate the reflection used to generate the respective dark fields.

dimension of 300 nm, this value is relatively coarse in size, and is attributed to the effect of temperature during the thermomechanical process. The presence of C_{60} further confirms the results from Figure 4, thus C_{60} is found in the sintered as well as in the thermomechanically processed samples. The tetragonal C_{60} is a unique phase that is observed in this work and has not been identified previously in the as alloyed or as sintered products. Although, this tetragonal structure has previously been reported in the as synthesized fullerene mix, and has been demonstrated to be resistant to mechanical milling [9,27]. It is important to mention that it is not well understood if the tetragonal C_{60} phase is synthesized during the SPS, the thermomechanical process or is rather a highly resistant phase. There is a possibility that this tetragonal C_{60} phase is present in the original fullerene mix and stands mechanical milling, SPS sintering and plastic deformation in severe (high temperature) environments; if this is the case, the tetragonal C_{60} is a highly resistance phase that requires further attention.

Figure 8 shows TEM images of the Fe-fullerene composite in the as thermomechanically processed condition with 70 % plastic deformation. A small crystal is shown in bright and dark field (Figs. 8a and 8b, respectively) containing a large number of nanostructured diamonds. The diamonds are characterized by TEM electron diffraction and a representative pattern is given in the insert of Figure 8b. The lattice parameter of the diamond particles is 0.36 nm that is <1 % different from the theoretical lattice parameter for cubic diamond [33]. The fullerene-diamond phase transformation has been previously reported for SPS processed carbon nanotubes, but in that case the carbon nanotubes have transformed at significantly higher temperatures (1773 K) [32]. The effect of pressure in a constrained condition most likely affects the temperature of transformation during rolling in the present investigation. Figure 8c shows a SEM image with a relatively large Fe_3C particle and its respective energy dispersive spectroscopy X-ray spectrum.

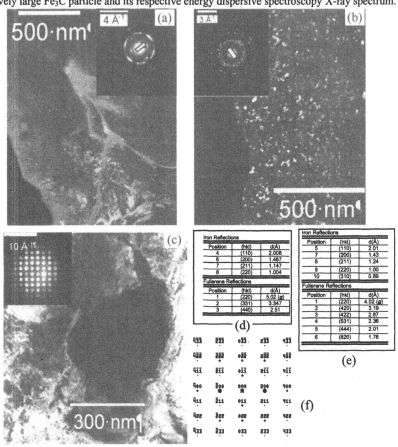

Iron Reflections (d)

Position	(hkl)	d(Å)
4	(110)	2.008
6	(200)	1.487
7	(211)	1.147
8	(220)	1.004

Fullerene Reflections

Position	(hkl)	d(Å)
1	(220)	5.02 (g)
2	(331)	3.347
3	(440)	2.51

Iron Reflections (e)

Position	(hkl)	d(Å)
5	(110)	2.01
7	(200)	1.43
8	(211)	1.24
9	(220)	1.00
10	(310)	0.89

Fullerene Reflections

Position	(hkl)	d(Å)
1	(220)	4.02 (g)
2	(420)	3.19
3	(422)	2.87
4	(531)	2.36
5	(444)	2.01
6	(820)	1.78

Figure 7. TEM image of particles identified in the Fe-fullerene sintered and thermo-mechanically processed composites with (a) 21% and (b-c) 43% deformation. Arrows indicate the reflection used to generate the respective dark fields. (d) – (e) Tables after indexing the diffraction patterns of (a) and (b) respectively. (f) Simulated diffraction pattern for tetragonal fullerene.

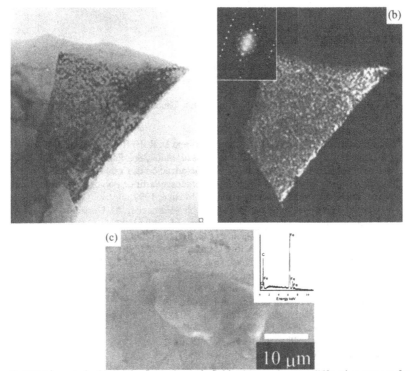

Figure 8. TEM images in (a) bright field, (b) dark field with an electron diffraction pattern for synthesized diamond and (c) SEM micrograph showing a Fe_3C particle and its respective energy dispersive X-ray spectrum. Specimens taken from the thermomechanically processed composite with 70 % deformation.

CONCLUSIONS

It is possible to manufacture a Fe-fullerene nanostructured composite by means mechanical alloying and SPS. The fullerene mix shows good control agent properties that prevent excessive agglomeration of the mechanically alloyed powders. C_{60} is a molecule that stands mechanical alloying in the presence of iron. The sintered composite shows a hardness of 722 μHV that is more than 700% higher than pure iron.

ACKNOWLEDGEMENTS

FCRH would like to express his gratitude to the University of Houston and the government of Texas for the start up package funding (HEAF .37215-B5008). The authors would like to thank Drs. M. Umemoto and V. Garibay Febles for their outstanding support with mechanical alloying and Spark Plasma Sintering methods. CONACYT and SIP-COFAA-IPN are acknowledged for financial support through grants 28952U/58133 (both authors) and SIP-20091310.

REFERENCES

1. P. J. F. Harris, 1st ed. (Cambridge, Cambridge University Press, 1999).
2. H.W. Kroto, J.R. Heath, S.C.O'Brien, R. F. Curl and R. E. Smalley, Nature. **318**, 162 (1985).
3. S. Iijima, Nature. **354**, 56 (1991).
4. Q. Ru, M. Okamoto, Y. Kondo, K. Takayanagi, Chem. Phys. Lett. **259**, 425, (1996).
5. D. Ugarte, Nature. **359**, 707 (1992).
6. H. Terrones, M. Terrones, J. Phys. Chem. Solids. **58**, 1789 (1997).
7. W. Krätschmer, L.D. Lamb, K. Fostiropoulos and D. R. Huffman, Nature. **347**, 354 (1990).
8. M. Umemoto, K. Masuyama and K. Raviprasad, Mater, Sci. Forum, **47**, 235 (1997).
9. F. C. Robles Hernandez, "Producción y Caracterización de Compósitos Metal-C (donde; Metal=Al o Fe y C=grafito o fullereno) Obtenidos a Partir de Polvos de Aleado Mecánico", MSc. Thesis, Instituto Politecnico Nacional, Mexico, 1999.
10. V. Garibay-Febles, H. A. Calderon, F. C. Robles-Hernández, M. Umemoto, K. Masuyama, J. G. Cabañas-Moreno, Mats. and Manufac. Proc. **15**, 547, (2000).
11. Z.G. Liu et al., J. of Phys. and Chem. of Sol. **61** 1119 (2000).
12. L. Díaz Barriga Arceo et al., J. Alloys Compd. **434–435**, 799 (2007).
13. L. Díaz Barriga-Arceo et al., J. Phys.: Condens. Matter, **16**, S2273 (2004).
14. J. S. Benjamin, Mater. Sci. Forum. **88-90**, 1 (1992).
15. P.S. Gilman and J.S. Benjamin, Ann. Rev. Mater. Sci. **13**, 279 (1983).
16. C. Suryanarayana, Prog. Mater Sci. **46**, 1, (2001).
17. J. Guerrero-Paz et al., Mats. Sci. Forum, **360-362**, 317 (2001).
18. J. H. Hafner, Chem. Phys. Lett. **296**, 195 (1998).
19. H. Dai, et. al., Chem. Phys. Lett. **260**, 471 (1996).
20. T. Gou, et. al., Chem. Phys. Lett. **243**, 49 (1995).
21. W. Zhou, et. al., Chem. Phys. Lett. **350**, 6 (2001).
22. Y. Gogotsi, N. Naguib, J. A. Libera, Chem. Phys. Lett. **365**, 354 (2002).
23. S. C. Tsang, Y. K. Chen, P. J. F. Harris, M. L. H. Green, Nature. **372**, 159 (1994).
24. S. C. Tsang, P. J. F. Harris, M. L. H. Green, Nature. **362**, 520 (1993).
25. L. Sun et al., Science. **312**, 1199, (2006).
26. L. Qian, et. al. Nano Lett., **8**, 4539 (2008).
27. F. C. Robles Hernandez, H. A. Calderon, under review, submitted August 2009.
28. P. M. Ajayan, J. M. Tous, Nature. **447**, 1066 (2007).
29. S. N. Kim, J. F. Rusling, F. Papadimitrakopoulos, Adv. Mater. **19**, 3214 (2007).
30. D. M. Hulbert, A. Anders, D. V. Dudina, J. Andersson, D. Jiang, C. Unuvar, U. Anselmi-Tamburini, E. J. Lavernia, A. K. Mukherjee, J. Appl. Phys. **104**, 033305 (2008).
31. D. M. Hulbert et al., Scripta Materialia. **60**, 835 (2009).
32. J. Shen, F. M. Zhang, J. F. Sun, Y. Q. Zhu, D. G. McCartney, Nanotechnol. **17**, 2187 (2006).
33. B. D. Cullity, Elements of X-Ray Diffraction, 1st Ed. (Asddison-Wesley Puublishing Company, Inc., United States of America, 1956).
34. M. Nikolussi et al., Scr. Mater. **59**, 814 (2008).

Mater. Res. Soc. Symp. Proc. Vol. 1243 © 2010 Materials Research Society

Heat Treatment Effect on Properties of Ni-P-Al$_2$O$_3$ Composite Coatings

Carlos A. León-Patiño, Josefina García-Guerra, Ena A. Aguilar-Reyes, José Lemus-Ruiz
Instituto de Investigaciones Metalúrgicas, Universidad Michoacana de San Nicolás de Hidalgo, A.P. 888 Centro, C.P. 58000, Morelia, Mexico

ABSTRACT

Ni-P and Ni-P-Al$_2$O$_3$ composite coatings are obtained by electroless plating on steel substrates. Alumina particles with an average particle size of 5 microns are added to the bath in loads of 5, 10, 15 and 20g/L. It is found a maximum retention of 18.2 vol.% Al$_2$O$_3$ for a ceramic load of 10g/L. The composition of the binary Ni-P deposits is 9.3 wt.% P and the balance nickel. The addition of ceramics to the electroless solution induces a reduction of phosphorous content to 9.0, 8.3, 7.9 and 7.3%, respectively. The deposited coatings are heat-treated in the temperature range between 100 and 500°C and holding times from 30 to 300 minutes. A maximum hardness of 1600 HV$_{0.1}$ is obtained for composite coatings containing 18.2 vol.% Al$_2$O$_3$ treated at 400 °C/1h. The uniform distribution of ceramics and precipitation of fine Ni$_3$P and Ni$_{12}$P$_5$ precipitates are responsible of the hardening of the nickel matrix.

INTRODUCTION

Electroless Ni-P deposits have been the subject of various investigations owing to their excellent properties, including hardness, wear resistance, corrosion resistance and magnetic characteristics. Since these properties depend primarily upon the structure and phosphorus content of the deposits, many studies have been carried out to characterize the as-deposited structure as a function of phosphorus content [1,2]. Notwithstanding the several applications of Ni-P deposits, their properties can be further enhanced by co-deposition of hard or lubricant particles such as SiC, Al$_2$O$_3$, B$_4$C and PTFE [3-7]. One of the main purposes of the ternary deposits is to improve the wear behavior of material surfaces; therefore, the fraction and nature of reinforcing particles, and composition and thermal history of the matrix are of important consideration. These deposits have a high hardness, particularly after heat treatment above 250°C [8]. Among the reinforcing ceramics, Al$_2$O$_3$ is extensively used due to its high hardness, elastic modulus, wear and good oxidation resistance. Despite several investigations, there is a little information on the effect of alumina content on the heat treatment response of Ni-P matrices, and its relationship with the phase evolution and hardness properties of Ni-P-Al$_2$O$_3$ composite coatings. In the present study, the microstructure and hardness properties of electroless Ni-P-Al$_2$O$_3$ with different amounts of reinforcing alumina is investigated after heat treatment in the temperature range 100-500°C.

EXPERIMENTAL DETAILS

Electroless Ni-P and Ni-P-Al$_2$O$_3$ coatings are applied on the surface of O1 tool steel substrates with dimensions 2x2x0.3 cm. The substrates are surface finished with silicon carbide abrasive paper to grade 600; then cleaned with acetone and NaOH solutions at room temperature and 50°C, respectively, finally they are activated in a 30 vol.% HCl solution for 2 min. The composition of the electroless bath is as follows: 30 g/L nickel chloride; 10 g/L sodium succinate; 10 g/L glycine; 2 ppm lead nitrate. The amount of the reducing agent, sodium hypophosphite, is varied from 4 to 83 g/L to obtain a high phosphorous Ni-P deposit, giving the results shown in Table I (ICP analysis, Termo Jarrell Ash AtomScan 16). The plating process is carried out at 90°C during 30 min and pH 5.0. The composite Ni-P-Al$_2$O$_3$ deposits are obtained by adding loads of 5, 10, 15 and 20 g/L of α-alumina particles (Norton materials, purity>99.4%, median size from the cumulative mass distribution D$_{50}$=5 μm) to the solution containing the maximum amount of sodium hypophosphite under continuous magnetic stirring.

Table I. Phosphorous content in the Ni-P deposits as a function of reducing agent

NaH$_2$PO$_2$·H$_2$O (g/L)	4	7	13	20	27	33	83
wt.% P	3.5	5.0	6.5	6.8	7.2	8.0	9.3

The crystallization behavior of the as-deposited amorphous coatings is followed through DSC analyses at a scanning rate of 10°C/min (SETARAM). On the basis of these results, coated samples are subjected to heat treatment from 100 to 500°C and a varying time in the range 30-300 min to study the phase evolution with temperature. The binary Ni-P and composite Ni-P-Al$_2$O$_3$ coatings are characterized by SEM (JEOL JSM-840A) and XRD (SIEMENS D5000) in the as-coated and heat-treated conditions. Vickers microhardness measurements (Zwick/Roell indentec ZHV) are performed using a pyramidal indenter with 100g load and 15s holding time.

RESULTS AND DISCUSSION

Characterization of as-deposited coatings

The same electroless formulation that leads to 9.3 wt.% P content in the Ni-P matrix, deposit called as Ni-9.3P in the rest of the paper, is selected for preparing the ternary Ni-P-Al$_2$O$_3$ composites in order to follow the variation of phosphorous concentration with the addition of alumina particles. The high phosphorous deposit is chosen because of Ni-P deposits containing ≥9.0 wt.% P are known to show interesting wear and corrosion properties. It is found that the addition of alumina particles into the electroless solution induces co-deposition of Al$_2$O$_3$ in the Ni-P matrix (Table 2). The weight percent of Al$_2$O$_3$ retained in the composite coatings is determined by a gravimetric method after completely removing the deposits by dissolution in nitric acid, centrifuging separation of ceramics and drying. It is found that 10 g/L of ceramic particles in the solution constitutes a critical load; this concentration of alumina in suspension provides the maximum incorporation of particles in the deposit (18.2 vol.% Al$_2$O$_3$). Larger additions decrease the retention of ceramics in the coating due to saturation and collision between particles near the steel substrate.

Not only retention of Al$_2$O$_3$ particles is affected with the addition of ceramics. It is found that the larger the concentration of alumina in suspension, the lower the phosphorous content in the deposit (Table 2). Reports in the literature suggest that the positive charge of the alumina surface attracts the H$_2$PO$_2^-$ ions in solution, reducing the availability of phosphorous for co-deposition [9]. The alumina particles embedded in the composite Ni-P-Al$_2$O$_3$ deposits disrupt the nodular structure reached in the binary Ni-9.3P coatings; however the composites remain continuous and uniform in thickness over the whole surface of the substrates. As observed in the cross section image of figure 1a, the coating applied is uniform, continuous and well adhered to

50

the surface of the substrate. The alumina particles embedded in the coating disrupt the usual nodular morphology formed in the surface of the Ni-P metallic deposits; however, as observed in figure 1b, the ceramics distribution is homogeneous over the whole surface of the substrates.

Table II. Al_2O_3 content and matrix composition of the Ni-P-Al_2O_3 composite coatings.

Al_2O_3 loaded to the electroless bath (g/L)	Al_2O_3 retained in the composite deposit (vol.%)	wt.% P in the Ni-P matrix	$HV_{0.1}$
(plain steel substrate)			245
0	0	9.3	435
5	8.3	9.0	745
10	18.2	8.3	814
15	14.6	7.9	766
20	10.1	7.3	735

 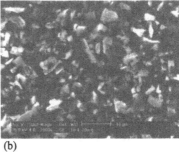

(a)   (b)

Figure 1. (a) Cross section and (b) surface morphology of a Ni-8.3P-18.2 vol.% Al_2O_3 deposit.

High phosphorous Ni-P coatings are normally amorphous. This condition is confirmed by the XRD glancing patterns of figure 2a. Irrespective of the alumina content in the composite deposits, the amount of phosphorous is always large enough to confer an amorphous character to the coatings. The broad peak in the 2θ region 40-50° is assigned to the amorphous state of the nickel-metalloid alloy. TEM observations performed on the deposits show diffraction rings of diffuse nature from selected-area diffraction (SAD) patterns confirming their amorphous nature (insert in Fig. 2b). The amorphous structure is due to a random deposition of phosphorous atoms that prevents the formation of tridimensional structures i.e., the nickel matrix is highly distorted giving non-periodic arrangements of atoms.

The amorphous deposits are heat-treated to induce crystallization and a likely improvement in hardness properties. DSC thermal analysis performed on Ni-9.3P deposit reveals two exothermic events in the range 230-350°C that suggest that crystallization occurs (Fig. 2b). This information confirms the amorphous character of the deposits and is useful for defining the thermal cycles for heat treatment.

Heat treatment and phase evolution

Table II presents the $HV_{0.1}$ hardness of the deposits in the as-coated condition. The values for the deposits are always higher than the hardness of the plain steel substrate.

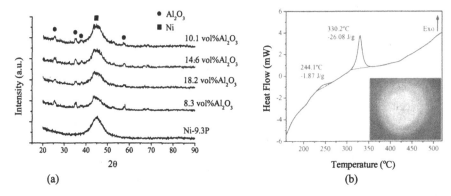

(a)   (b)

Figure 2. (a) XRD profiles of as-coated Ni-9.3P and Ni-P-Al₂O₃ deposits. (b) DSC analysis of the amorphous Ni-9.3P deposit showing the crystallization behavior (the insert shows a SAD pattern in TEM of the amorphous Ni-9.3P deposit).

As previously explained, the addition of ceramics to the electroless bath not only affects the retention of alumina particles in the deposits, it also modifies the phosphorous content in the Ni-P matrix. Thus, hardness of the deposits is affected by both, P concentration and Al₂O₃ content. However, under the present conditions the variation in hardness is mainly due to the alumina particles embedded in the deposits. Generally speaking, it is found that the higher the co-deposited alumina, the higher the hardness value. A more important effect of Ni:P ratio in hardness properties is reported after heat treatment of deposits when Ni-P precipitates are formed [3-7].

Figure 3a shows the phase evolution of Ni-9.3P deposits after heat treatment at 300°C. Comparing these results with the XRD profile of the same sample in the as-coated conditions (Fig. 2), it is clear that the original broad peak of the amorphous structure becomes partially crystalline after 30 min treatment, as reveals the Ni (111) reflection formed. The sample treated during 60 min does not present significant changes; however after 120 min, there is evidence of precipitation of Ni₁₂P₅ and Ni₃P phases in a crystalline Ni (111) matrix. After 300 min, the deposit is completely crystalline; the matrix consists of nickel and fine Ni₃P precipitates which almost completely replace the metastable Ni₁₂P₅ phase observed at 120 min.

The XRD profiles of the heat-treated Ni-P-Al₂O₃ deposits at 300°C/60 min are shown in Figure 3b for the different volume fractions of Al₂O₃ added. Generally speaking, crystallization of the amorphous deposits takes place in shorter periods of time; they also consist of crystalline nickel and Ni₃P and Ni₁₂P₅ precipitates. It is considered that the presence of alumina particles reduces the incubation time for precipitation of Ni-P particles. The ceramic surfaces act as nucleation sites reducing the energy barrier for the process to proceed. The crystallization phenomenon is accompanied by strengthen of the alumina-matrix interfaces due to a reduction of porosity and atomic arrangement of the initial amorphous phase.

Hardness of heat-treated deposits

Generally speaking the hardness of heat-treated Ni-9.3P deposits increases with aging time and temperature (Fig. 4a). However hardening of the nickel matrix is not possible even for a long exposure time at 100°C and 200°C due to a reduced precipitation of Ni-P phases. At higher temperatures >300°C, the crystallization kinetics allows the formation of crystalline nickel (Fig. 3) and the deposits become hard. A maximum hardness is reached at 400°C after 60 min, the ensuing reduction of hardness is most likely due to the growth of Ni_3P precipitates and nickel grains. At 500°C a similar overaging effect is registered, with a consequent decrease in hardness.

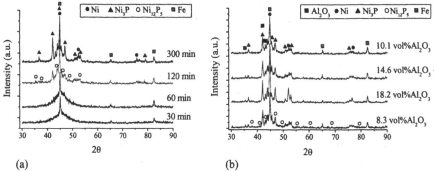

(a)　　　　　　　　　　　　　　　(b)

Figure 3. XRD profiles of heat-treated deposits. (a) Binary Ni-9.3P at 300°C as a function of time; (b) Composite Ni-P-Al$_2$O$_3$ at 300°C/60 min and different Al$_2$O$_3$ contents.

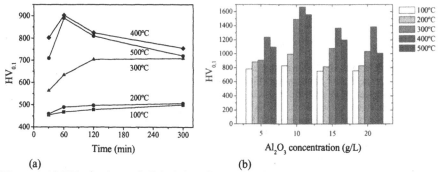

(a)　　　　　　　　　　　　　　　(b)

Figure 4. (a) HV$_{0.1}$ hardness of Ni-9.3P deposits as a function of time and temperature, (b) HV$_{0.1}$ hardness of heat-treated Ni-P-Al$_2$O$_3$ composites as a function of temperature and Al$_2$O$_3$ content in the electroless bath.

In the case of the Ni-P-Al$_2$O$_3$ deposits, the alumina particles behave like a reinforcing phase that uniformly distributes the stresses in the coating and inhibits the free movement of dislocations. The mechanical strength of the deposits is thus improved i.e, the hardness of the composite coatings is higher with respect to the binary Ni-9.3P deposit. As can be seen in figure

53

4b, the maximum hardness values are developed with an Al_2O_3 concentration of 10 g/L which corresponds with the maximum retention of 18.2 vol. % Al_2O_3. As in the case of the binary Ni-9.3 deposits, heat treating at 400 °C induces considerable hardening. The microhardness for deposits treated at 400 °C reach values twice as large as those at 100 °C. Higher temperatures produce an overaging effect that leads to a slight reduction in hardness.

CONCLUSIONS

Uniform and continuous Ni-9.3P coating and composite Ni-P-Al_2O_3 deposits are prepared by electroless plating. A critical load of 10 g/L of ceramics is found, giving a maximum incorporation of 18.2 vol%Al_2O_3 in the deposit; larger additions decrease the retention of ceramics in the coating due to saturation and collision effects. Phosphorous content decreases in the composite deposits with the concentration of alumina particles in suspension. Irrespective of the matrix composition and ceramics content, the deposits show an amorphous structure in the as-coated conditions; after heat treatment, crystalline nickel and fine Ni_3P precipitates are formed. According to XRD results, the crystallization of composite coatings is relatively faster since alumina surfaces behave as external nucleus for Ni-P precipitates formation. The uniform distribution of ceramics and precipitation of fine Ni_3P and $Ni_{12}P_5$ phases are responsible of the hardening of the nickel matrix.

ACKNOWLEDGMENTS

The financial support by CONACYT-Mexico (Grant 57299) and COECYT-Michoacan (Grant 08-02/132) is acknowledged. Josefina García-Guerra also thanks CONACYT-Mexico by the scholarship provided.

REFERENCES

1. W. Riedel, *Electroless Nickel Plating*, (ASM International Finishing Publications, 1991) p.3-55, 81-131.
2. P. Sampth Kumar and P. Kesavan Nair, J. Mat Sci Lett. 13 671 (1994).
3. J.N. Balaraju, T.S.N. Sankara Narayanan and S.K. Seshadri, 33 807 (2003).
4. Ming-Der Ger, Mater. Chem. Phys. 76 38 (2002).
5. Y.L. Shi, Mater. Chem. Phys. 87 154 (2004).
6. S. Alirezaei, S.M. Monirvaghefi, M. Salehi and A. Saatchi, Surf. Coat. Tech. 184 170 (2004).
7. J.N. Balaraju, K.S. Kalavati, K.S. Rajam, Surf. Coat. Tech. 200 3933 (2006).
8. V.D. Papachristos, C.N. Panagopoulos, U. Wahlstrom, L.W. Christoffersen and P. Leisner, Mater. Sci. Eng. A 279 217 (2000).
9. A. Abdel Aal, Z.I. Zaki and Z. Abdel Hamid, Mater. Sci. Eng. A 447 87 (2007).

Mater. Res. Soc. Symp. Proc. Vol. 1243 © 2010 Materials Research Society

Thermochemical Method for Coating AISI 316L Stainless Steel with Ti

Jorge López-Cuevas[1], José L. Camacho-Martínez[2], Juan C. Rendón-Angeles[1], Martín I. Pech-Canul[1] and Juan Méndez-Nonell[1,3]
[1]CINVESTAV-IPN Unidad Saltillo, Ramos Arizpe, 25900 Coah., México
[2]CIATEQ, Querétaro, 76150 Qro., México
[3]CIQA, Saltillo, 25253 Coah., México

ABSTRACT

This paper presents a thermochemical method, based on a mixture of molten alkaline halides to produce a Ti coating on AISI 316L stainless steel. The thickness of the coatings is a function of temperature and time. It is observed that the physical form of the Ti source employed affects both coating thickness and morphology. The formation of several inter-diffusion layers is detected, each having a characteristic chemical composition, morphology and location at the substrate/coating interface. It is proposed that some of the produced Ti coatings can be employed to improve osseointegration of stainless steel for potential prosthetic devices.

INTRODUCTION

In the late 1960s, Cook and co-workers developed the so-called "Metalliding" process, which involves the diffusion of atoms of one metal into the surface of another, giving rise to the formation of an alloyed surface. This process occurs when metals are immersed in a molten fluoride electrolytic bath [1]. A broad range of materials are produced with new surface properties, such as improved resistance to oxidation, corrosion and wear. The metalliding process is based on the formation of a galvanic cell. When the anode is electrically connected with the cathode by an external connector, a REDOX reaction occurs and the anode is oxidized producing metallic ions and electrons. The metallic ions are dissolved into the molten bath and diffuse toward the cathode surface while electrons flow through the external connector toward the same electrode. At the cathode, the incoming metallic ions are reduced by the flow of electrons to atoms of the anode metal, which then diffuse into the surface of the electrode. This process is self-supported by a galvanic cell electromotive force and continues as long as there is diffusion. Thus, no external electrical potential is needed; although an external electric current may be imposed in order to accelerate the process. Metalliding has been studied more recently by Cardarelli et al. [2] for the formation of tantalum films on stainless steels. Okabe and co-workers [3,4] have studied this processes in more detail and rename it as Electronically Mediated Reaction (EMR). Two possible EMR modes are recognized. In Short Range EMR (SR-EMR), a homogeneous reaction takes place in an electronically conducting molten salt and leads to the formation of metallic powders dispersed in the bath. In contrast, in Long Range EMR (LR-EMR) a heterogeneous reaction takes place and electron transfer occurs through an electronic conductor (reactor wall, stirrer, or the metallic deposit itself). In this case, the morphology of the deposit can be dendritic, spongy, etc. The mechanism of formation of metallic coatings is essentially the same for metalliding and LR-EMR: formation of a galvanic cell. In the case of SR-EMR, however, the acting mechanism seems to depend on the system studied. For instance, when this process is applied to coat iron substrates with Ti, a reaction of Ti^{2+} to metallic Ti and Ti^{3+} seems to occur at the substrate surface that leads to production of loose Ti powder and Ti coatings. This method can also be used to coat ceramic substrates with Ti [5]. N.C. Cook [1], Straumanis et al. [6], Steinman et al. [7], and several other researchers, have studied this processes. In their work, a substrate and a Ti source (powder, sponge or thin foil) are commonly introduced into a bath of molten salts, generally consisting of eutectic mixtures of alkaline halogenides with or without additions of K_2TiF_6. Initially, it was believed that corrosion of Ti by the molten salts resulted in formation of a colloidal suspension (a pyrosol), which was in turn responsible for the coating of the substrate. At present, it is known [8] that Ti reacts with the molten salts in the bath giving rise to formation of $TiCl_2$, $TiCl_3$ and $TiCl_4$. It has been determined [9] that, at equilibrium, four disproportionated heterogeneous chemical reactions take place simultaneously:

$$2Ti^{3+} + Ti \leftrightarrow 3Ti^{2+} \qquad (1)$$
$$2Ti^{3+} \leftrightarrow Ti^{4+} + Ti^{2+} \qquad (2)$$
$$4Ti^{3+} \leftrightarrow Ti + 3Ti^{4+} \qquad (3)$$
$$2Ti^{2+} \leftrightarrow Ti + Ti^{4+} \qquad (4)$$

In these reactions, Ti is in solid state while Ti^{4+} is in the form of gaseous $TiCl_4$ in equilibrium with soluble $TiCl_6^{2-}$ complex. Reactions 1 and 2 occur predominantly in the presence of a Ti excess with large surface area, originating the formation of a clear solution rich in Ti^{2+}. Significant losses of Ti occur due to the low stability of Ti^{4+} in the bath. Under these conditions, reactions 3 and 4 take place at any solid surface making contact with the molten salts, such as crucible walls and substrate surface, which then become coated with Ti. The two main disadvantages of this technique are the $TiCl_4$ losses and a possible re-dissolution of the Ti coating (reaction 1) [10]. The thickness and microstructure of the Ti coating depend on the duration of the process, bath temperature and composition, including the concentration of Ti, as well as on working atmosphere and oxidation degree of the Ti source [11,12].

The main objective of this work is to develop a metalliding method to produce a Ti coating on the surface of AISI 316L stainless steel [13,14] using an eutectic mixture of molten NaCl-KCl salts.

EXPERIMENTAL PROCEDURE

An eutectic 43.94 wt% NaCl-56.06 wt% KCl mixture (melting point of ~ 654°C) is prepared using high purity (~99.9%) KCl and NaCl. Ti thin foil (with a thickness of 0.25 mm), Ti sponge (with a particle size smaller than 1.2 mm) and Ti powder (with a particle size smaller than 150 μm) are used to produce the coatings. Purity of these materials is in the range 99.7 to 99.95%. The substrates are cylindrical pieces of commercially available cold-rolled AISI 316L stainless steel, with a diameter of 2.54 cm and a height of ~0.65 cm. The circular flat end faces of the substrates are ground and polished with alumina with a particle size of 0.05μm. The samples are washed, rinsed, degreased, dried and stored in a dessicator until they are used. The chemical composition of the AISI 316L stainless steel is (wt.%): 0.027C, 0.47Si, 1.51Mn, 0.016P, 0.024S, 16.71Cr, 2.71Mo, 10.05Ni, 0.47Cu, 0.049V and Fe. The experiments are carried out during 4-400 minutes at 750-1000 °C, under a flow of ultra high purity Ar, using an especially adapted electric furnace. The experimental setup is shown in Figure 1.

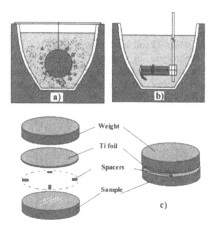

Figure 1. Sample arrangements used for (a) Ti powder or sponge and (b) Ti thin foil; (c) detailed sandwich-type configuration shown in (b); the spacers are made of Ti-foil with a thickness of ~0.25 mm; the counterweights are AISI 316L Stainless Steel cylinders similar in size to the coated samples.

Three simultaneous experiments are conducted for each set of conditions studied. In the first sample configuration, Fig. 1a, prior to the experiments the samples are placed in 100 ml porcelain crucibles and covered with a mixture of 100 g of salt and 5 g of either Ti powder or sponge. In the second sample configuration, Fig. 1b, the sandwich-type arrangement is placed in the crucible and covered with pure salt mixture. K-type thermocouples placed inside the molten salt baths and in several furnace locations are used for temperature monitoring and control. At the end of the experiments the furnace is switched off allowing the samples to cool down in a natural way covered with the salt mixture, maintaining the Ar flow. Finally, the samples are extracted from the furnace, washed and rinsed with hot deionized water and dried. The surface of the coatings is analyzed by X-Ray Diffraction (Philips 3040) using Ni-filtered CuK$_\alpha$ radiation. The microstructure of the coatings is characterized in cross-sections by optical (Olympus Vanox AHMT3) and scanning electron microscopy (Jeol JSM 6300 SEM /Noran EDS system). The samples are mounted in non-conductive bakelite and sectioned at a plane perpendicular to the flat faces. After that, they are prepared using standard metallography techniques and, finally, coated with graphite for their analysis. Concentration profiles for Fe, Ni, Cr, Mo, Mn and Ti are determined at the substrate/Ti coating interface. Coating thicknesses are measured by Image Analysis (Image Pro Plus software) using digital images acquired on the SEM. Lastly, both the stainless steel grain size (ASTM E112) and Rockwell B hardness (Wilson/Rockwell B504-T) are determined before and after each experiment.

DISCUSSION
Coating thickness and morphology

Ti coatings with thicknesses ranging from <1 to 100 μm are obtained. Regardless of the Ti source employed, the coating thickness increases with increasing temperature. However, Ti foil produces much thicker coatings than Ti powder or sponge. On the other hand, the coating morphology is very regular when Ti powder or sponge is used, Fig.2a, but it is very irregular (exhibiting dendritic growth) when Ti foil is employed, Fig.2b. This kind of growth has been observed for the case of electrolytic Ti coatings [15] and has been attributed to a high concentration of this metal in the molten salt bath. In this work, a high concentration of Ti is produced locally due to the very short distance between the Ti foil and the substrate surface when using the sandwich-type configuration.

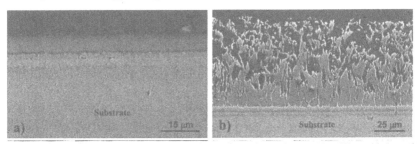

Figure 2. SEM-BEI images of Ti coatings obtained at: a) 430 min/800°C (Ti powder), and b) 100 min/850°C (Ti thin foil).

In contrast, Ti powder and sponge tend to sediment to the bottom of the crucible since they have a larger density than the molten salt. As a result, the distance between Ti-source and substrate surface is larger. This, in turn, increases the time required for the diffusion of Ti, lowering simultaneously its concentration at the substrate/bath interface. This results in the formation of relatively thin and regular non-dendritic Ti coatings. This is further promoted by a lack of mechanical stirring of the bath. Since dendritic growth implies the presence of a considerable amount of open porosity in Ti coatings, as can be seen in Fig. 2b, the occurrence of this phenomenon can be used in order to improve the osseointegration properties of stainless steel implants. Porous implants made either of Ti [16,17] or Ti-Ni alloy [18] have shown excellent adhesion properties to bone.

Interdiffusion phenomena occurring at the coating/substrate interface

Thickening of Ti coatings as a function of time and/or temperature indicates that this is a diffusive process. In fact, the chemical elements contained in the substrate tended to diffuse into the coating while Ti tended to diffuse in the opposite direction towards the stainless steel matrix. This gave rise to formation of several inter-diffusion layers (Fig. 3a), particularly in samples treated at the highest temperature during the longest time. Concentration profiles determined at the substrate/coating interface for the main alloying elements of AISI 316L stainless steel, as well as for Ti, are shown in Fig. 3b.

The phase composition of the layers observed in Fig. 3a is determined from the SEM/EDS results and it is given in Table I. As can be seen, layer 1 is a Ti-rich Ti-Fe-Ni solid solution. The composition of this layer is consistent with the β-Ti phase reported by Ghosh and Chatterjee [19]. Some needles of NiTi and NiTi$_2$ phases can also be found at the β-Ti grain boundaries,

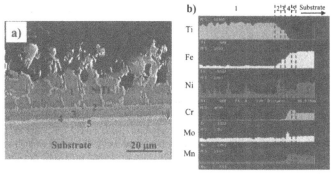

Figure 3. a) SEM-BEI micrograph of Ti coating (64 min/950°C, Ti thin foil), showing the different diffusive layers formed; b) Concentration profiles determined for the main diffusing elements (64 min/1000°C, Ti thin foil).

Table I. Results of SEM/EDS analysis of interfacial zone.

Layer	Composition (wt. %)					Assumed phase composition	Atomic ratio for the assumed phases	
	Fe	Cr	Ni	Mo	Ti		Theoretical	Analyzed
1	13.6	1.36	4.4	---	79.74	*Ti-Fe-Ni solid solution (β-Ti)	**N.A.	**N.A.
2	33.49	2.93	6.8	---	55.15	Ti₂Fe	Ti/Fe = 2	Ti/Fe = 1.92
3	37.15	3.45	7.43	---	50.97	TiFe + Ti₂Fe	Ti/Fe = 1 for TiFe	Ti/Fe = 1.60
4	60.7	24.10	3.56	3.56	7.43	σ phase	***(Fe,Ni)/(Cr,Mo) = 1.5	(Fe,Ni)/(Cr,Mo) = 2.3
5	63.73	23.55	4.40	2.41	3.56	σ phase	As above	(Fe,Ni)/(Cr,Mo) = 2.5

*NiTi and NiTi₂ phases are also present in this layer.
**N.A. = Not Applicable. ***Parentheses indicate solid solutions.

growing normally to the substrate surface. This is similar to the results obtained by Kundu and Chatterjee [20] in experiments conducted for 1h at 930°C, using a Ni interlayer between Ti and 304 stainless steel. They find the presence of discrete β-Ti "islands" surrounded by a NiTi₂ matrix at the Ni-Ti interface. It is also deduced that layers 2 and 3 correspond to TiFe intermetallic phases. Layers 4 and 5 are similar in composition to the σ phase reported by Ghosh and Chatterjee [19], and thus, they are assumed to correspond to this phase. In addition to σ phase, these authors also observed β-Ti and TiFe phases similar to those found in the present material. They also note an accumulation of Cr at their σ phase zone, which is attributed to a decrease in the activity of this metal in the steel matrix due to the diffusion of Ti into it. For this reason, Cr diffusion takes place in the direction of decreasing activity gradient, instead of occurring in the direction of decreasing concentration gradient. In this work, this behavior is observed not only for Cr, but also for Mo and Mn, at the σ phase zones, which suggests that a similar decrease in activity takes place for the latter two metals at the steel matrix, due to the diffusion of Ti into this matrix.

CONCLUSIONS

The thickness of Ti coatings produced by a metalliding process in a mixture of molten NaCl-KCl salts increases with increasing temperature. This effect is more pronounced when a Ti thin foil is used as a Ti source. The morphology of the coatings depends on the concentration of Ti in the molten salt bath. In general, five different inter-diffusion layers are observed at the substrate/coating interface, involving the formation of β-Ti, NiTi, TiFe and σ phases. It is proposed that the dendritic growth of Ti coating produced using a thin foil as a Ti source can be employed to improve osseointegration of AISI 316L stainless steel prosthetic devices.

REFERENCES

1. N.C. Cook, *Sci. Amer.* **221**, 38 (1969).
2. F. Cardarelli, P. Taxil, and A. Savall, *Int. J. Refract. Met. H.* **14**, 365 (1996).
3. T.H. Okabe, and Y. Waseda, *JOM-J. Min. Met. Mat. S.* **49**, 28 (1997).
4. T. Uda, T.H. Okabe, Y. Waseda, and Y. Awakura, *Sci. Technol. Adv. Mat.* **7**, 490 (2006).
5. P. Wei, H. Qiliang, C. Jian, C. Juan, and Y. Huang, *Mater. Lett.* **31**, 317 (1997).
6. M.E. Straumanis, S.T. Shin, and A.W. Schlechten, *J. Electrochem. Soc.* **104**, 17 (1957).
7. J.B. Steinman, R.V. Warnock, C.G. Root, and A.R. Stetson, *J. Electrochem. Soc.* **114**, 1018 (1967).
8. C.-H. Li, H.-B. Lu, W.-H. Xiong, and X. Chen, *Surf. Coat. Tech.* **150**, 163 (2002).
9. E. Chassaing, F. Basile, and G. Lorthioir, *J. Appl. Electrochem.* **11**, 187 (1981).
10. C.E. Baumgartner, *Anal. Chem.* **64**, 2001 (1992).
11. S.T. Shih, M.E. Straumanis, and A.W. Schlechten, *J. Electrochem. Soc.* **103**, 395 (1955).
12. A.W. Schlechten, M.E. Straumanis, and C.B. Gill, *J. Electrochem. Soc.* **102**, 81 (1955).
13. G. Vargas, M. Méndez, J. Méndez, and A. Salinas, U.S. Patent 5,482,731 (1996).
14. M.H. Fathi, M. Salehi, A. Saatchi, V. Mortazavi, and S.B. Moosavi, *Dent. Mater.* **19**, 188 (2003).
15. I.A. Menzies, D.L. Hill, G.J. Hills, L. Young, and J. O'M. Bockris, *J. Electroanal. Chem.* **1**, 161 (1959).
16. K. Asaoka, N. Kuwayama, O. Okuno, and I. Miura, *J. Biomed. Mater. Res.* **19**, 699 (1985).
17. M. Thieme, K.P. Wieters, F. Bergner, D. Scharnweber, H. Worch, J. Ndop, T.J. Kim, and W. Grill, *J. Mater. Sci. - Mater. M.* **12**, 225 (2001).
18. B.-Y. Li, L.-J. Rong, Y.-Y. Li, and V.E. Gjunter, *Intermetallics* **8**, 881 (2000).
19. M. Ghosh, and S. Chatterjee, *Mater. Sci. Eng. A* **358**, 152 (2003).
20. S. Kundu, and S. Chatterjee, *Mater. Sci. Eng. A* **425**, 107 (2006).

Mater. Res. Soc. Symp. Proc. Vol. 1243 © 2010 Materials Research Society

Titanium Coatings on AISI 316L Stainless Steel Formed by Thermal Decomposition of TiH₂ in Vacuum

Jorge López-Cuevas[1], José L. Rodríguez-Galicia[1], Juan C. Rendón-Angeles[1], Martín I. Pech-Canul[1] and Juan Méndez-Nonell[1,2]
[1]CINVESTAV-IPN Unidad Saltillo, Ramos Arizpe, 25900 Coah., México
[2]CIQA, Saltillo, 25253 Coah., México

ABSTRACT

Ti-coated AISI 316L stainless steel, for potential biomedical applications, is obtained by thermal decomposition of TiH₂ under vacuum. The presence of hydrogen in the coating material facilitates the sintering process of Ti particles, with simultaneous formation of several inter-diffusion layers at the substrate/coating interface, whose thickness and chemical composition depend mainly on the treatment temperature. Coatings prepared at 1100°C exhibit formation of a wide zone at the substrate/coating interface, which is associated with the appearance of cracks, and which consists of a mixture of $\lambda + \chi + \alpha$-Fe phases. Formation of abundant microporosity is also observed in this region, which is attributed to the Kinkerdall effect.

INTRODUCTION

TiH₂ dissociates into hydrogen and metallic Ti at ~525°C under vacuum [1]. It contains ~96 wt % Ti and is inert to water, air and most acids [2]. During its thermal dissociation, hydrogen migrates into the furnace atmosphere, while metallic Ti covers the container walls with a thin film. This phenomenon has been employed since the 1950's in order to produce Ti coatings at the surface of some ceramic materials such as alumina. After being coated with Ti, the ceramic pieces can be better wetted by molten metals during the preparation of metal-matrix composites, or can be joined more easily with other ceramic or metallic materials. Other applications of TiH₂ include the formation of aluminum foams [3] and hydrogen storage [2]. In the latter case, hydrogen in Zn or Fe solid solutions can be easily liberated to feed internal combustion engines or fuel cells. Thermal dissociation of TiH₂ under vacuum has not been reported previously for the production of Ti coatings on the surface of ferrous or non-ferrous metals or alloys. Thus, in the present work this method is used for the formation of Ti coatings on the surface of AISI 316L stainless steel, for potential biomedical applications. However, although the studied method constitutes an easy way to potentially increase the biocompatibility and osseointegration properties of AISI 316L Stainless Steel [4, 5], further work would have to be carried out in order to address several related important issues that may arise. First, the necessity to limit the formation and growth, at the substrate/coating interface, of any phases and microporosity that can be deleterious for the material's mechanical strength [6]. Next, there could be some biocompatibility concerns if the resulting Ti coatings have a relatively high Ni concentration (although it is known [7] that some Ti-Ni alloys are biocompatible, especially when the formation of a film of TiO₂ is promoted at their surface [8]). The risk of a likely occurrence of galvanic corrosion in the case of formation of Ti-Fe solid solutions in the coatings [9], and, lastly, the likely precipitation of σ phase and $M_{23}C_6$ carbides in the AISI 316L stainless steel due to the production of Ti coatings at high temperatures, which in turn would have a deleterious effect on the mechanical properties and corrosion resistance of the substrates [10].

EXPERIMENTAL DETAILS

Reagent grade TiH$_2$, with a mean particle size of 20 μm, is employed. The substrates are cylindrical pieces of commercially available cold-rolled AISI 316L stainless steel, with a diameter of 2.54 cm and a height of ~0.65 cm, and with a chemical composition (wt %) of 0.027%C, 0.47%Si, 1.51%Mn, 0.016%P, 0.024%S, 16.71%Cr, 2.71%Mo, 10.05%Ni, 0.47%Cu, 0.049%V and Fe (balance). The circular flat faces of the substrates are prepared metallographicaly, with a final polish using alumina with a particle size of 0.05 μm. Then, the samples are washed, rinsed, degreased, dried and stored in a desiccator prior to cover their polished surfaces with a paint prepared with 25 g of TiH$_2$ in a solution of ~1.5 g of an acrylic resin (acryloid B48N, Rhom and Haas, added as binder) in 100 ml of reagent grade amyl acetate. Then, the samples are dried in hot air and placed inside rectangular ceramic containers, avoiding touching the container walls with the painted surfaces. The samples are treated for 30 or 120 minutes at 900 or 1100°C, under a vacuum of ~1-3x10^{-3} mbar, using a Thermolyne 59300 fused quartz tube furnace (with a length of 90 cm and internal diameter of 50 mm), equipped with a rotary vacuum pump (Leybold S25B). The sample temperature has been monitored by a K-type thermocouple touching the ceramic container. Each experiment is conducted three times. At the end of the experiments, the furnace is switched off, allowing the samples to cool down to room temperature under vacuum. Then, the samples are extracted from the furnace and the surface of the Ti coatings are analyzed by X-Ray Diffraction (Philips 3040), using Ni-filtered Cu K$_\alpha$ radiation. The surfaces are prepared metallographicaly and coated with graphite for analysis by Scanning Electron Microscopy (Jeol JSM 6300 SEM /Noran EDS system). Concentration profiles of Fe, Ni, Cr, Mo, Mn and Ti are determined at the substrate/coating interface. Coating thicknesses are measured by Image Analysis (Image Pro Plus software), using digital images acquired on the SEM.

DISCUSSION

Sintering of Ti coatings

Partial sintering of the coatings is observed at 900°C. At this temperature, an increased degree of Ti particle consolidation and a decreased porosity are observed in the coatings when time is increased from 30 to 120 minutes (Figs.1a and 1b). This is associated with a progressive reduction in the thickness of the coatings. In contrast, at 1100°C, the Ti particle sintering process concludes after 30 min, Fig. 1c, with a minimum amount of porosity remaining after this time. The production of porous pieces of Ti for dental implants has been reported in the literature by using a two-step process involving a pre-sintering stage at 1000°C, followed by sintering at 1400°C for 24h [11]. Thieme et al. [12] have also reported a two-step process (1150°C/1h + 1470°C/1h) for the production of porous Ti orthopedic implants. Thus, temperatures much higher than those used in the present work are usually employed for the sintering of Ti powder, which suggests that in our case there is a factor facilitating the occurrence of this process. Such factor could be the hydrogen contained in TiH$_2$. It is known [13] that when Ti powder is sintered in combination with a reducing agent, the sintering temperature can be decreased to 1000°C. It is also known that the removal of hydrogen during Ti sintering accelerates this process. Senkov and Froes [14] have mentioned that sintering of hydrogenated Ti occurs more easily, compared with non-hydrogenated Ti, suggesting that hydrogen cleans up the powder surface, promoting a better bonding between adjacent particles. In the present case, further evidence that the Ti sintering is facilitated by hydrogen contained in the coating material comes from the observation of a minimal amount of residual porosity remaining after this process, which is much lower than the 35-55% residual porosity that normally results after pressureless sintering of Ti powder [15]. Regarding the starting point of the Ti powder sintering, Fraval and Godfrey [16] indicate that Ti powder with minimum sintering can be obtained by dehydrogenation of TiH_2 below 700°C. These observations, combined with the present results, indicate that it is possible to obtain fully or partially sintered Ti coatings on AISI 316L stainless steel by choosing a suitable temperature in the range of 700-1100°C for the thermal decomposition of TiH_2 under vacuum.

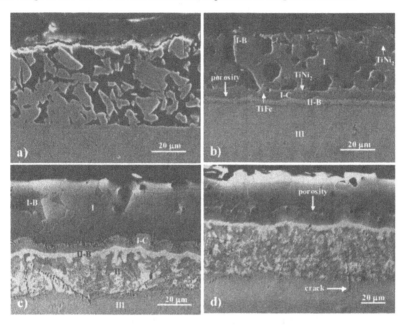

Figure 1. SEM-BEI images of Ti coatings obtained after 30 or 120 min at 900°C (a and b), and at 1100°C (c and d).

Formation of interdiffusion zones at the coating/substrate interface

In general, the alloying elements of the stainless steel, namely Fe, Ni, Cr and Mo, tend to migrate gradually into the coating, while Ti tends to diffuse in the opposite direction, towards the steel matrix. The penetration distance in the latter direction is shorter than in the Ti coating, especially in samples treated at 900°C. Ghosh and Chatterjee [17] observe a similar situation after 90 min of diffusion bonding of Ti on 304 Stainless Steel at 950°C. This is attributed to the fact that the fcc structure of austenite is more compact than the crystalline structure of Ti. These considerations apply also to our experimental results, since AISI 316L stainless steel has also a fully austenitic matrix (Fig. 2a).

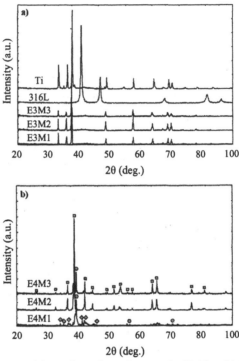

Figure 2. XRD patterns of the surface of samples coated with Ti at (a) 900°C/120 min. [all reflections for 316L stainless steel correspond to austenite (JCPDS card 33-0397) and for Ti-coated EM samples correspond to α-Ti (JCPDS card 44-1294)], and at (b) 1100°C/120 min. [□ = NiTi₂ (JCPDS card 72-0442), ○ = TiFe (JCPDS card 19-0636), and ◊ = TiFe₂ [18]].

Although not all of them are detected by XRD (Fig. 2a), probably due to their small thicknesses, the SEM/EDS analysis reveals the formation of four different inter-diffusion zones in the samples treated for 120 min at 900 °C. These zones are identified as I, I-B, I-C and II-B in Fig. 1b. Zone III is the unaffected region of the substrate. Zones I and I-B are Ti rich, with a content of ~98 wt % of this metal in the latter zone, while zone I contains 85.1 wt % Ti, 10.9% Fe, 1.8% Ni, 1.3% Cr and 0.4% Mo. Based on the results reported by Ghosh and Chatterjee [17], it is assumed that both zones correspond to β-Ti and α-Ti phases, respectively. A few NiTi₂ intermetallic needles are also found at the boundary between I and I-B regions (right upper zone in Fig. 1b). On the other hand, since zone I-C contains 51.5 wt % Ti, 38.1% Fe, 5% Ni, 4 % Cr, 1% Mn and 0.5% Mo, it is deduced that this layer is likely constituted by a mixture of NiTi₂ and TiFe intermetallic phases. Zone II-B contains 58.1 wt % Fe, 26.1% Cr, 6.8% Ti, 4% Ni, 3.1% Mo and 2% Mn. Again, based on the results of Ghosh and Chatterjee [17], zone II-B is attributed to the formation of σ phase. It is important to point out that this zone contains more Cr and Mo than both substrate (zone III) and zone I-C, as can be clearly seen in Fig. 3a. Ghosh and

64

Chatterjee [17] observe a similar situation at their σ phase zone for the case of Cr, which is attributed to a decrease in the activity of this metal at the steel matrix due to the diffusion of Ti into it. The stainless steel employed by these researchers is practically Mo free, however, based on our results, it can be said that in our case the activity of Mo is also decreased by diffusion of Ti into the steel matrix. It is also evident from Fig. 3a that the concentration of Ni is higher in zone I-C than in zones I and II-B, mainly near the boundary with zone I. Lastly, Fig. 1b shows an accumulation of microporosity which occurred at the boundary between zones I-C and II-B, and which is probably due to the so-called Kirkendall effect, since it is known [19] that the diffusion rate of Ni in Ti is faster than that of Ti in Ni. The location of this microporosity probably coincided with the position of the Kirkendall plane.

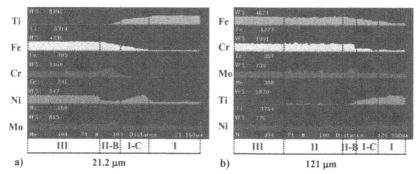

Figure 3. Line scan analyses carried out by SEM/EDS for samples coated with Ti at (a) 900°C/30 min and (b) 1100°C/120 min.

Again, XRD is limited to detect and identify all the phases formed at the substrate/coating interface (Fig. 2b) in the materials treated at 1100°C. According to the SEM/EDS analysis, in the coatings produced at 1100°C/30 min (Fig. 1c), the I-B phase appears in the form of discrete "islands" surrounded by phase I. This analysis indicates that most likely the latter two phases correspond to β-Ti. Under these conditions, Zone I-C is thicker than the corresponding region in samples treated at 900°C, showing a high concentration of Ti and Ni at the boundary with zone I (Fig. 3b). Again, this layer is likely constituted by a mixture of $NiTi_2$ and TiFe intermetallic phases. It can also be observed that the II-B phase becomes thicker, and that a new and wide region II, not observed at 900°C, appears underneath it. Region II contains 65.3 wt % Fe, 17.4% Cr, 6.3% Ni, 5.7% Ti, 2.9% Mo and 1.9% Mn. Based on the findings of Kundu and Chatterjee [20], who studied the diffusion bonding of Ti on 304 stainless steel, employing a Ni interlayer between both materials, it is assumed that at 1100°C the II and II-B zones are constituted by mixtures of λ + χ + α-Fe and λ + α-Fe phases, respectively. It is important to note that the formation of these two layers is not detected in the same system by Ghosh and Chatterjee [17], when the Ni interlayer is not employed. They observe instead the formation of σ phase, as already mentioned. Kundu and Chatterjee [20] point out that the Ni interlayer inhibits the diffusion of Ti toward the steel matrix up to 900°C. In this work, a similar situation can arise at all treatment temperatures in use, due to the accumulation of Mo at layer II-B, which can act as a barrier for the diffusion of Cr toward the coating and of Ti toward the steel matrix, but not for the diffusion of Fe and Ni in the latter direction. This results in the formation of λ, χ and α-Fe phases in zone II at 1100°C. A marked accumulation of microporosity observed at this zone is likely due to the Kirkendall effect. Furthermore, the presence of cracks in zone II is probably due to differences in the thermal expansion coefficients and mechanical properties of its constituent phases. Lastly, it is observed that the coating thickness is simply a function of the amount of TiH_2 paint applied to the sample surface, and that the extension of the inter-diffusion zones depends mainly on the treatment temperature.

CONCLUSIONS

Ti particles generated by the thermal dissociation of TiH_2 under vacuum are fully sintered at ~1000°C at the surface of the AISI 316L Stainless Steel samples. This is attributed to a likely cleaning effect of hydrogen contained in the TiH_2. In the 900-1100°C temperature range, diffusion of Fe, Cr, Ni and Mo takes place from the stainless steel side toward the Ti coating, with simultaneous diffusion of Ti in the opposite direction. This gives rise to the appearance of several inter-diffusion zones, involving the formation of various phases, which in turn are associated with the appearance of cracks at the coating/substrate interface. Lastly, an accumulation of microporosity is observed at this interface, which is likely due to the occurrence of the Kinkerdall effect.

REFERENCES

1. I. Gotman, and E.Y. Gutmanas, *J. Mater. Sci. Lett.* **9**, 813 (1990).
2. M.A. Klochko, and E.J. Casey, *J. Power Sources* **2**, 201 (1978).
3. J. Banhart, *Int. J. Vehicle Des.* **37**, 114 (2005).
4. G. Vargas, M. Méndez, J. Méndez, and A. Salinas, U.S. Patent 5, 482,731 (1996).
5. M.H. Fathi, M. Salehi, A. Saatchi, V. Mortazavi, and S.B. Moosavi, *Dent. Mater.* **19**, 188 (2003).
6. P. He, J. Zhang, R. Zhou, and X. Li, *Mater. Charact.* **43**, 287 (1999).
7. B.-Y. Li, L.-J. Rong, Y.-Y. Li, and V.E. Gjunter, *Intermetallics* **8**, 881 (2000).
8. J.-X. Liu, D.-Z. Yang, F. Shi, and Y.-J. Cai, *Thin Solid Films* **429**, 225 (2003).
9. M.E. El-Dahshan, A.M. Shams El Din, and H.H. Haggag, *Desalination* **142**, 161 (2002).
10. D.N. Wasnik, G.K. Dey, V. Kain, and I. Samajdar, *Scripta Mater.* **49**, 135 (2003).
11. K. Asaoka, N. Kuwayama, O. Okuno, and I. Miura, *J. Biomed. Mater. Res.* **19**, 699 (1985).
12. M. Thieme, K.P. Wieters, F. Bergner, D. Scharnweber, H. Worch, J. Ndop, T.J. Kim, and W. Grill, *J. Mater. Sci. - Mater. M.* **12**, 225 (2001).
13. Z.S. Rak, and J. Walter, *J. Mater. Process. Tech.* **175**, 358 (2006).
14. O.N. Senkov, and F.H. Froes, *Int. J. Hydrogen Energ.* **24**, 565 (1999).
15. O. Takashi, O. Tadashi, W. Munetoshi, and K. Masamichi, Jpn. Application Patent WO2002JP01332 20020215 (2002).
16. J.T. Fraval, and M.T. Godfrey, U.S. Patent 6,475,428 (2002).
17. M. Ghosh, and S. Chatterjee, *Mater. Sci. Eng. A* **358**, 152 (2003).
18. T. Suzuki, T. Saikusa, H. Suematu, W. Jiang, and K. Yatsui, *Surf. Coat. Tech.* **169-170**, 491 (2003).
19. D. Tomus, K. Tsuchiya, M. Inuzuka, M. Sasaki, D. Imai, T. Ohmori, and M. Umemoto, *Scripta Mater.* **48**, 489 (2003).
20. S. Kundu, and S. Chatterjee, *Mater. Sci. Eng. A* **425**, 107 (2006).

Mater. Res. Soc. Symp. Proc. Vol. 1243 © 2010 Materials Research Society

Effect of Annealing on the Magnetic Properties of a Cold Rolled Non-Oriented Grain Electrical Steels

N.M. López G.[1] and A. Salinas R.[2]

Centro de Investigación y de Estudios Avanzados del Instituto Politécnico Nacional, Saltillo Campus, P.O. Box 663, Saltillo Coahuila, México 25900. e-mails: [1]nan_mar0904@hotmail.com, [2]armando.salinas@cinvestav.edu.mx

ABSTRACT

The effect of plastic deformation and subsequent annealing on the microstructure and magnetic properties (hysteresis core losses) of non-oriented grain semi-processed Si-Al electrical steel sheet are investigated. Plastic deformation of strip samples is performed by cold-rolling (5-20% reduction in thickness) along the original rolling direction. Annealing is carried out in air during 1 or 60 minutes at temperatures between 650 and 850°C. Measurements of B-H hysteresis curves are performed using a Vibrating Sample Magnetometer and characterization of annealed microstructures is carried out using optical metallography. The results show that hysteresis losses increase by a factor between 1.2 and 2.0 as the magnitude of the applied plastic deformation increases from 5 to 20% reduction in thickness. The rate of recovery of energy losses as a result of annealing depends on annealing time. Short annealing times produce full recovery of the effect of cold work and values of energy losses lower than in undeformed material. The magnitude of the additional recovery increases with strain but does not depend on annealing temperature. Long annealing times, which induce complete recrystallization, and either normal or abnormal grain growth, enhance recovery of hysteresis losses. The rate of recovery increases as both the strain and annealing temperature increase. Recovery of the deformation microstructure and internal stress relief produce only limited recovery of the magnetic properties. However, recrystallization and grain growth brings about a significant decrease in hysteresis losses.

INTRODUCTION

Plastic deformation of semi-processed or processed electrical steel strips gives rise to a significant increase in magnetic energy losses [1-6]. Therefore, it is of great technological interest to have a full understanding of the origin of this effect, as well as in finding optimum annealing conditions that allow recovery of the strip's original magnetic properties. Theoretically, coercivity is found to be proportional to the square root of the dislocation density [7] and dislocation density in iron increases linearly with strain [8]. This behavior - an increase of coercive force and magnetic losses with the square root of strain - has been found experimentally by some authors [4, 5]. However, Astié et al. [9] defined three stages of magnetic hardening associated with changes of the dislocation structure and residual stresses in pure iron. Deformations below 2% produce only isolated dislocations and, as a result, no change in coercivity is observed. In contrast, a large increase in coercivity occurs when dislocation tangles are formed at strains between 10% and 15%. Finally, the rate of increase in coercive force decreases when a dislocation cell substructure is developed at strains larger than 15%. These microstructure changes and the resulting internal stresses are usually held responsible for magnetic domain wall pinning [9-12]. It has been suggested that 90° domain walls are more sensitive to the stress field of dislocations than 180° domain walls [9, 10, 13], which are more sensitive to the pinning by grain boundaries and dislocation cell walls.

When a plastically deformed electrical steel strip is subjected to an annealing treatment its magnetic properties are recovered. However, little has been published about microstructural and magnetic changes due to annealing, particularly for the case where annealing conditions produce only microstructural recovery, i.e., a decrease of the dislocation density and formation of subgrain boundaries. This paper presents the results of an investigation on the effect of annealing on the magnetic properties of semi-processed Si-Al electrical steel sheets deformed by cold rolling from 5 to 20% reduction in thickness.

EXPERIMENTATION

A 0.57 mm thick, non-oriented grain electrical steel sheet is obtained from a local supplier. The chemical composition of the strip is determined by optical emission spectrometry (wt%C=0.03, wt%Si=0.383, wt%Al=0.217, wt%Mn=0.594, wt%S<0.0005, wt%P=0.043, wt%Cr=0.053, wt%Mo=0.018, wt%Fe=balance). Samples 280 mm long and 150 mm wide are cut from the strip and cold rolled to 5, 10% and 20% reduction in thickness. The rolling direction is parallel to the original rolling direction of the strip. Annealing of the cold rolled samples is performed during either 1 or 60 minutes at temperatures between 650 and 850°C under an air atmosphere in a muffle-type furnace. Magnetic hysteresis loops are determined at 1.2 T and 60 Hz, with the applied field parallel to the rolling direction, using a Vibrating Sample Magnetometer (Lake Shore Series 7300 VSM System). The changes in the microstructure produced by deformation and annealing are determined by standard optical metallography techniques using a reflected light microscope (Olympus Vanox AHMT3).

RESULTS AND DISCUSSION

Effect of plastic deformation on the microstructure and magnetic properties.

Figure 1 illustrates the effect of plastic deformation (cold rolling) on the microstructure of the strip samples. As can be seen, regardless of the applied deformation, the microstructures consist of approximately equiaxed grains with an average grain size of about 21 μm). The effects of the magnitude of the applied plastic deformation on energy losses due to hysteresis are shown in Figs. 1. As can be seen, hysteresis losses increase by a factor between 1.2 and 2.0 as the magnitude of the applied plastic deformation increases from 5 to 20% reduction in thickness. The rate of increase is low up to about 7% strain and then increases rapidly.

Assuming that the crystal orientation distribution in the electrical steel investigated is random, each grain contains a number of 180° domain walls (DW) aligned along easy magnetization directions and closed at their edges by 90° DWs [14]. Movement of these magnetic domains is required for magnetization and demagnetization of the material and any obstacle to this movement produces changes in magnetic behavior. In the present material, the magnitude of the plastic deformation applied by cold rolling did not have any apparent effect on the grain microstructure (Fig. 1). However, analysis of the X-ray diffraction patterns of deformed samples showed an increase of the FWHM for the (110) and (222) diffraction peaks (Fig. 2a).

Since the size of the grains in the deformed samples is rather large, the increase in FWHM with increasing deformation can be attributed, at least in principle, to increase in the dislocation density and non-uniform deformation of grains with different orientations. This observation suggests that changes in the dislocation structure of deformed samples may be responsible for the observed increase in energy losses. Moreover, Fig. 2b shows that the coercivity increases rapidly with the applied strain up to ~10% and then the rate of increase with increasing strain is significantly reduced. A similar result has been reported by Astié et al. [6], in pure Fe. These authors suggested that the initial increase in coercivity is due to formation of dislocation tangles and that, at larger strains, the rate of increase in coercive force decreases when a dislocation cell substructure is developed. A similar behavior of the coercivity is observed in Fig. 2 for the present material. Thus, it appears that creation of dislocations by plastic deformation reduces the dimensions across which magnetic domain movement can occur and this results in a rapid increase in energy losses for rolling strains larger than about 7% (Fig. 1).

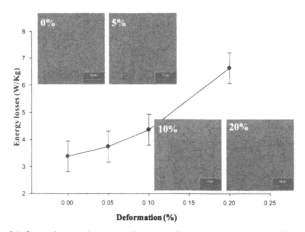

Fig.1. Effect of deformation on the magnetic properties and microstructure of semi-processed, non-oriented grain electrical steel strip cold rolled to (a) 0%, (b) 5%, (c) 10%, (d) 20% reduction in thickness.

Recovery of magnetic properties by annealing.

Quantitative metallography of the microstructures produced after deformation and annealing during 1 minute at temperatures between 650 and 850 °C revealed only limited effects on grain size; the average grain size varied between 20 and 28 μm. As an example, Fig. 3a illustrates the effect of prior strain on the microstructure after annealing during 1 minute at 850°C. Nevertheless, as shown in Fig. 3b, within this grain size range, the increase in energy losses observed as a result of cold work (Fig. 1) is fully reverted and the energy losses due to hysteresis reach values of about 3.1 W/kg, i.e. about 10% lower than those observed in the as received material. These results indicate that, even for material deformed to 20% strain by cold rolling and annealed during 1 minute at temperatures as high as 850 °C, recrystallization and grain growth of the deformed microstructures is rather difficult. Therefore, the effect of the short

69

annealing treatment on the energy losses due to hysteresis (Fig. 3b) is mainly due to recovery processes such as dislocation reconfiguration (polygonization) and relief of internal strain energy.

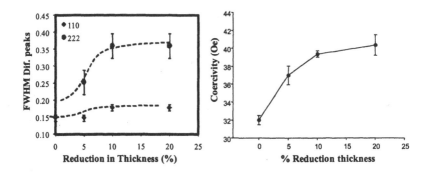

Fig.2. Effect plastic deformation by rolling on: a) the breadth (FWHM) of the (111) and (110) XRD peaks and b) the coercivity of semi-processed, non-oriented grain electrical steel strip.

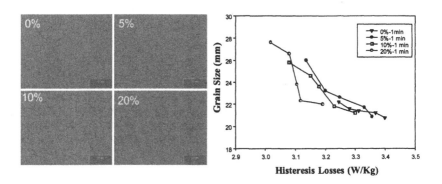

Fig. 3 a) Effect of prior strain on the microstructure after annealing during 1 minute at 850°C, b) Effect of grain size (produced by cold rolling to the indicated reductions in thickness and annealing during 1 minute at temperatures between 650 and 850 °C) on the energy losses due to hysteresis.

In contrast, as show in Fig. 4a, annealing during 60 minutes at 850 °C resulted in recrystallization by strain-induced grain boundary migration and abnormal grain growth even in undeformed samples. In this case, the effect of prior strain and temperature on the average grain size of the annealed samples is much more pronounced. Annealing temperatures from 650 to 800 °C produced average grain sizes below 40 µm independently of the combination of prior strain and annealing temperature. However, it is observed that the rate of increase of average grain size with temperature up to about 800 °C increases with applied strain. After that, the rate of increase in grain size becomes independent of the prior deformation. For a given annealing temperature the final average grain size after a 60 minute annealing increases considerably with increasing deformation reaching maximum values between 60 and 120 µm when annealing is carried out at 850 °C (Fig. 4a). The effect of grain size on the energy losses due to hysteresis of samples annealed during 60 minutes is illustrated in Fig. 4b. As can be seen, the energy losses decrease rapidly with increasing grain size and the minimum losses that can be obtained after a 60 minute annealing depends strongly on the magnitude of the applied strain. In the case of the material deformed to 20% reduction in thickness and annealed during 60 minutes at 850 °C, the energy losses are about 2.3 W/kg, i.e. about 30% lower than in the as received material.

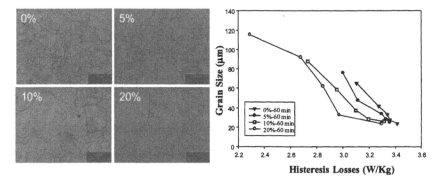

Fig.4. a) Effect of prior strain on the microstructure after annealing during 60 minutes at 850°C, b) Effect of grain size (produced by cold rolling to the indicated reductions in thickness and annealing during 60 minutes at temperatures between 650 and 850 °C) on the energy losses due to hysteresis.

Campos et al. [15] attributed a large increase of hysteresis losses due to cold-rolling of 0.5% Si electrical steel sheets to the development of large residual microstresses (150–350 MPa). The present results suggest that relief of these residual microstresses takes place very rapidly during the initial stages of annealing at temperatures between 650 and 850 °C and that the recovery of the magnetic losses does not depend on the level of applied deformation.

It is well known that grain boundaries are strong pinning sites for magnetic domain wall (DW) movement. The present results show that, in the present material, decreasing the grain boundary area by recrystallization and grain growth decreases significantly the energy losses due to hysteresis. The magnitude of the decrease in energy losses depends first, on the amount of applied deformation an, second, on the temperature and time of annealing. Thus, in the present material, grain size controls the energy losses due to hysteresis.

CONCLUSIONS

1. Plastic deformation by cold rolling of annealed non-oriented grain Si-Al electric steel strips gives rise to an important increase in energy losses (by a factor of 1.2 to 2) as the strain is increased from 5 to 20%. This effect is attributed to the dislocation structure resulting from the applied deformation which has an influence on magnetic domain wall movement during magnetization.

2. The rate of recovery of energy losses as a result of annealing depends on annealing time and temperature as well as on the applied deformation. Short annealing times give rise to a complete recovery of the effect of cold work and a further recovery that increases with strain but does not depend on annealing temperature. Long annealing times, which produce complete recrystallization and grain growth, enhance recovery of hysteresis losses. In this case, the rate of recovery increases with both the magnitude of the applied strain and the annealing temperature and time.

REFERENCES

[1] O. Hubert, M. Clavel, I. Guillot, E. Hug, J. Phys. IV France 9 (1999)207.
[2] F.J.G. Landgraf, M. Emura, J. Magn. Magn. Mater. 242 (2002) 152.
[3] M.J. Sablik, F.J.G. Landgraf, R. Magnabosco, M. Fukuhara, M.F. de Campos, R. Machado, F.P. Missell, J. Magn. Magn. Mater. 304(2006) 155.
[4] L.J. Swartzendruber, G.E. Hicho, H.D. Chopra, S.D. Leigh, G. Adam, E. Tsory, J. Appl. Phys. 81 (1997) 4263.
[5] C.K. Hou, IEEE Trans. Magnetics 30 (1994) 212.
[6] F.J.G. Landgraf, M. Emura, K. Ito, P.S.G. Carvalho, J. Magn. Magn. Mater. 215 (2001) 94.
[7] M. Kersten, Z. Angew, Phys. 8 (1956) 496.
[8] A.S. Keh, S. Weissmann, Electron Microscopy and the Strength of Crystals, Interscience, New York, 1963.
[9] B. Astié, J. Degauque, J.L. Porteseil, R. Vergne, IEEE Trans Magnetics 17 (1981) 2929.
[10] H. Trauble, in: Berkowitz, Kneller (Eds.), Magnetism and Metallurgy, Academic Press, New York, 1969, p. 622.
[11] O. Hubert, E. Hug, I. Guillot, M. Clavel, J. Phys. IV France 8 (1998) Pr2–515.
[12] O. Hubert, L. Hirsinger, E. Hug, J. Magn. Magn. Mater. 196–197 (1999) 322.
[14] J.M. Makar, B.K. Tanner, J. Magn. Magn. Mater. 184(1998) 193.
[15] R.M Bozorth, Ferromagnetism, 1st ed., D. Van Nostrand, Toronto, 1951.
[16] M. de Campos, M.J Sablik, F.J.G Landgraf. Journal of Magnetism and Magnetics Materials. 320 (2008).

Mater. Res. Soc. Symp. Proc. Vol. 1243 © 2010 Materials Research Society

Effect of Tensile Deformation on the Grain Size of Annealed Grain Non-Oriented Electrical Steel

J. Salinas B., A. Salinas.

Centro de Investigación y Estudios Avanzados del IPN. P. O. Box 663, Saltillo Coahuila, México 25000.
e-mails: jorgesb231182@yahoo.com.mx[1], armando.salinas@cinvestav.edu.mx[2]

ABSTRACT

An experimental study on the effect of tensile deformation on recrystallized grain size has been carried out in order to establishing the optimal deformation needed to accelerate grain growth during final annealing of semi-processed non-oriented Si-Al, low C electrical steel sheets. The material is deformed in tension to strains from 3 to 20% and then air-annealed at temperatures between 700 and 900 °C. The results show that the critical deformation for recrystallization (8%) is independent of annealing temperature. However, the critical recrystallized grain size increases with annealing temperature from 160 to 240 μm. After that, the grain size decreases exponentially with increasing deformation. Abnormal grain growth is observed in samples annealed at 700 °C after strains in the range from 7 to 12%. This type of behavior is also observed in specimens annealed at 800 and 900 °C, however, in this case the pre-strain range is expanded to 3-12%. Normal grain growth is observed in samples pre-deformed to strains larger than 12%. In this case, the final grain size after 2 hour anneal is about 55 μm, also independent of annealing temperature. The possible implications of these results on the magnetic properties of these materials are discussed.

INTRODUCTION

Electrical steels are processed by following routes similar to those used for most cold rolled steels, i.e. hot rolling, cold rolling, box annealing, and temper rolling [1].
Grain non-oriented electrical steels are classified in two categories: semi-processed and fully-processed. This classification is based on an annealing treatment conducted after punching the steel sheets into the required shapes. In case of semi-processed grades, the final annealing is carried out at the customer's plant. On other hand, fully processed electrical steels are given the final annealing treatment after temper rolling at the steelmaker plant.

Temper rolling is the key step in manufacturing semi-processed electrical strips because the magnetic properties are greatly improved by strain induced boundary migration (SIBM) [2]. During the final annealing treatment after temper rolling (small deformations) SIMB [3,4,5] occurs leading to abnormal grain growth.

The recrystallized grain size varies with deformation before the final annealing treatment [6,7]. However, there exists a critical deformation which produces the coarsest grain size after annealing. This critical deformation usually is small [1,7]. The aim of this investigation is to determine the effect of tensile deformation on the recrystallized grain size of annealed, grain non-oriented semi-processed Si-Al steel. This information is important since it can be used to optimize temper rolling deformation to enhance grain growth during the final annealing of the strips after punching.

EXPERIMENTAL PROCEDURE

The chemical composition of the steel strip used in this study is shown in Table I. The strip is obtained from the production line of a local manufacturer after box annealing prior to temper rolling. The thickness of the strip is 0.580 mm and had a microstructure consisting of equiaxed ferrite grains with an average size of 17 μm. Tensile samples are machined from the strip with tensile axes parallel to the rolling direction. These samples are deformed to various amounts of elongation ranging from 3 to 20%. After that, the deformed samples are air-annealed at temperatures in the range of 700–900 °C during times from 30 to 7200 seconds.

The microstructures of annealed specimens are characterized on longitudinal cross-sections using optical microscopy. The samples are polished and subsequently etched in 2% Nital solution in order to reveal all the grain boundaries. Average grain size is evaluated using commercial image analysis software.

Table I. Chemical composition of the material

C	S	Si	Mn	Al	Cu	Mo	P	Ni
0.072	0.0012	0.570	0.577	0.211	0.034	0.014	0.042	0.033

RESULTS

The effect of tensile deformation on the grain size obtained after annealing during 7200 seconds at various temperatures is shown in Fig. 1. As can be seen, there is a strong deformation and annealing temperature dependence on the grain size. Annealing at 800 and 900 °C causes a very rapid increase in grain size reaching a maximum at about 8% tensile strain. In contrast, annealing at 700 °C produces no recrystallization in the sample deformed to a strain of 3%. Once the tensile strain is higher than 3%, the grain size increases in a similar form to that observed at the higher annealing temperatures, again reaching a maximum at 8% strain. This suggests that the critical strain for grain growth in the present material is 8% and does not depend on annealing temperature. However, the critical grain size does increase as the annealing temperature is increased.

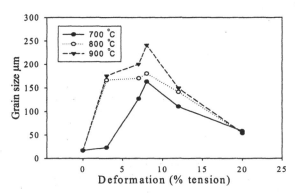

Figure 1. Effect of deformation and annealing temperature (7200 seconds) on the final grain size of grain non-oriented semi-processed Si-Al steel.

Fig.2 shows the microstructure of the sample deformed to 8% tensile strain and annealed at 700 °C during 1800 seconds. As can be seen, abnormal grain growth occurs both near the surface and in the interior of the strip. Similar results are observed in samples deformed 7% and 12%. In contrast, the sample deformed to 20 % strain exhibits uniform smaller grain size which does not depend on annealing temperature (see Fig. 1).

Figure 2. Abnormal grain growth in material strained by 8%, and annealed at 700 °C during 1800 seconds of treatment.

The evolution of the average grain size with annealing time at 700 °C is shown in Fig. 3 for different levels of tensile strain. As can be seen, there are three different types of behavior. First, for small deformations (3% or less) the driving force for recrystallization is too small and the grain size after annealing remains nearly constant up to 7200 s. Second, for strains between 8 and 12%, abnormal grain growth by strain-induced boundary migration of only a few grains (Fig. 2) takes place after an incubation period which decreases as the amount of deformation increases. After that, these grains continuously grow until they consume the deformed microstructure and reach a final constant size that decreases with the applied strain. Finally, for strains higher than 12%, recrystallization starts from the beginning of the annealing and no incubation period is observed in the grain size-time curve.

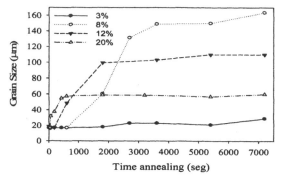

Figure 3. Effect of tensile deformation on evolution of grain size during isothermal annealing at 700 °C.

75

The annealing treatments carried out at 800 and 900 °C in samples deformed to strains larger than 12% produced recrystallization by nucleation and normal growth of new grains in of deformed microstructue. However, in these cases, preferential columnar grain growth is observed from the surface of the samples deformed at strains lower than about 12%. An example of this microstructure is illustrated in Figs. 4a and 4b for the sample strained 12% and annealed 600 and 5400 seconds, respectively, at 900 °C. These observations suggest that some of the surface grains grow preferentially until they consume the partially recrystallized microstructure during the annealing. As shown in Fig. 4c, preferential columnar grain growth did not occur in the sample deformed 20% and annealed at 900 °C. In this case, the final microstructure consists of smaller equiaxed grains.

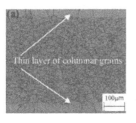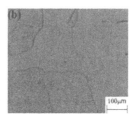

Figure 4. Grain morphology obtained after different annealing treatments. (a) sample strained 12% and annealed 600 s at 900°C, (b) sample strained 12% and annealed 5400 seconds at 900 °C, (c) sample strained 20% and annealed 7200 seconds at 900°C.

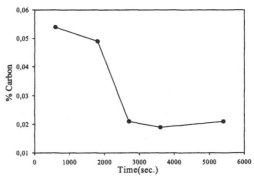

Figure 5. Effect of annealing time at 900 °C on the C content of the strip deformed to 12% tensile strain. C content is determined by the LECO combustion technique.

DISCUSSION

The final grain size after annealing of a deformed metallic material strongly depends on the magnitude of the applied strain. This effect indicates that the number of nuclei or the nucleation rate for recrystallization is affected by strain [8]. Accordingly, a higher strain will provide more nuclei per unit volume and hence a smaller final grain size. On the contrary, a smaller deformation produces less nuclei leading to larger grains as long as the applied deformation is sufficient to drive the nucleation of the new grains. In general, this type of recrystallization can only occur when the applied strain is larger than a certain critical value and the recrystallized grain size decreases exponentially with increased deformation. As shown in Fig. 1, this is not the case for the grain non-oriented electrical steel investigated in the present work. In this case, the grain size increases rapidly with applied strain reaching a maximum at about 8% tensile strain, and then decreases exponentially with increasing strain. This behavior can be explained in terms of a different type of recrystallization mechanism.

Recrystallization at small deformations can proceed by strain-induced boundary migration (SIBM) of deformed grains without nucleation of new grains [2, 5]. The driving force for this kind of recrystallization is the difference in stored energy between two neighboring grains and also depends on the differences in grain orientations [2, 3]. The stored energy due to small deformations is insufficient to produce nucleation of new recrystallized grains and only the grains with the lowest stored energy grow at the expense of grains with higher stored energy [6, 9]. As shown in Fig. 2, under conditions of small applied strains and low annealing temperatures, SIBM results in abnormal grain growth and may be used to improve magnetic properties of semi-processed non-oriented electrical steels. Magnetic properties of this type of steels are strongly dependent on grain size [2].

Data presented in Fig. 3 show that during isothermal annealing grain growth in the present material exhibits different kinetic behavior depending on the applied strain, i.e. no growth, abnormal growth and equilibrium growth. These results are similar to those found by other researchers [7, 10, 11] that proposed that polygonization is an important factor for abnormal growth in pure iron. In samples deformed to strains between 3 and 12 % and annealed at temperatures below Ar_1, grain growth exhibits an incubation period that is inversely proportional to strain. According to Antonione et al. [1], under these conditions, polygonization acts as an inhibitor for grain growth by reducing the driven force for boundary migration. Therefore, grain growth is only observed in grains with the largest differences in stored energy which leads to abnormal grain growth. In the case of material strained to 3%, deformation is too small to cause sufficiently large differences in stored energy or grain orientations to drive boundary migration and recrystallization by SIBM is not observed (Fig. 3). On the other hand, in material deformed to strains larger than 12%, the stored energy of most deformed grains is sufficient to cause nucleation of new recrystallized grains and, therefore, no abnormal grain growth is observed.

Preferential columnar grain growth is observed in samples deformed to small strains ($\leq$12%) and annealed at 800 and 900 °C (Fig. 4). It has been shown [1, 12] that development of columnar grain microstructures requires decarburization of the samples at temperatures in the range between A_1 and A_3. Figure 5 shows the effect of annealing time on the carbon content of samples deformed 12% and annealed at 900 °C. As can be seen, rapid decarburization of the strip occurred and, after 2700 s, the C content reached a constant value of 0.02%. During annealing under these conditions, austenite and ferrite coexist and abnormal grain growth in the bulk of the material cannot occur because austenite acts as pinning obstacles on ferrite grains [1, 13]. However, during annealing of the ferrite+austenite surface microstructure, decarburization causes destabilization of the austenite grains which then transform to ferrite, probably by growth of pre-existing ferrite [13, 14]. After decarburization is complete, the surface ferrite grains growth inwards due to their geometric a advantage (i.e. larger size) which results in the columnar grain morphology shown in Fig. 4b.

In the case of the sample deformed to a strain of 20% and then annealed at 900°C, the driving force for nucleation of new grains in the deformed microstructure is larger and causes recrystallization *before* decarburization is significant. Thus, columnar growth of surface ferrite grains does not take place.

CONCLUSIONS

Recrystallization in grain non-oriented electrical steel deformed in tension up to 12% strain proceeds by strain-induced boundary migration of deformed grains and results in abnormal grain growth for annealing temperatures between 700 and 900 °C. Annealing at 700 °C of samples deformed to strains lower than 3% show no recrystallization. In contrast, recrystallization in samples deformed to strains higher than 12% takes place by nucleation of new grains from the deformed microstructure.

Simultaneous recrystallization and decarburization of samples previously deformed to strains of 12% or less and annealed at temperatures between A_1 and A_3, leads to the development of columnar ferrite grain microstructures. This type of microstructure is formed by preferential growth of larger surface ferrite grains. It is suggested that these larger surface ferrite grains are formed by growth of pre-existing ferrite grains due to rapid decarburization of austenite grains at the surface of the strip.

REFERENCES

1. A. R. Mader, Metall. Trans. 17A, 1277 (1986).
2. Jongtae Park, Jerzy A. Szpunar and Sangyun Cha, Mater. Sci. Forum Vol. 408-412, 1263 (2002).
3. L. Kestens, J.J. Jonas, P.Van Houtte, and E. Aernoudt, Metall. Trans. 27A, 2347 (1996).
4. S. W. Cheong, E.J. Hilinski, and A.D. Rollett, Metall. Trans 34A, 1321 (2003).
5. F. J. Humphreys, Mater. Sci. Forum Vol. 467-470, 107 (2004).
6. K. Murakami, J. Tarasiuk, H. Regle and B. Bacroix: Mater. Sci. Forum Vol. 467-470, 893 (2004).
7. R. W. Ashbrook, Jr. and A.R. Mader: Metall. Trans. 16A 897 (1985).
8. F. J. Humphreys and M. Hartherly: *Recrystallization and Related Annealing Phenomena*, second edition (Elsevier Science, UK 2004) p.248.
9. Seung- Hyun Hong and Dong Nyung Lee, Mater. Sci. and Eng. A 375, 75 (2003).
10. C. Antonione, G. Della Gatta, G. Riontino, G. Venturello, J. Mater. Sci. 8, 1 (1973).
11. C. Antonione, F. Marino, G. Riontino, M. C. Tabasso, J. Mater. Sci. 12, 747 (1977).
12. F. Kovac, M. Dzubinsky, Y. Sidor, J. Magn. Magn. Mater. 296, 333 (2004).
13. Yuriy Sidor, Frantisek Kovac, Mater. Charac. 55 1 (2005).
14. Toshiro T. and Takashi T. ISIJ. 35, 548 (1995).

Mater. Res. Soc. Symp. Proc. Vol. 1243 © 2010 Materials Research Society

Influence of Cyclodextrin in the Synthesis of Magnetite

L. A. Cobos Cruz[1], C. A. Martínez Perez[1], A Martínez Villafañe[2], J. A. Matutes Aquino[2], J. R. Farias Macilla[2] and P. E. García Casillas[1]
[1] Instituto de Ing. y Tecnología, Universidad Autónoma de Cd. Juárez, Cd. Juárez Chih., México
[2] Centro de Investigaciones en Materiales Avanzados, Chihuahua, Chih., México

ABSTRACT

Cyclodextrin (CD) has been studied intensively due to its ability to form inclusion complexes with a variety of guest molecules in the solid state. A few studies have paid attention to the use of CD to facilitate the synthesis of inorganic nanoparticles. In this work the synthesis of magnetite (M) is made in the presence of CD. The particle size of the inorganic material is controlled by the presence of CD, in which spherical particles of few nanometers are grown. The synthesis of Fe_3O_4 (M) in the presence of α-cyclodextrin (α-CD) and β-cyclodextrin (βCD) is described. The formation of an M-CD complex is studied in both cases by Fourier transform infrared spectroscopy (FT-IR) in order to elucidate the chemical bonding of the complex. The morphology and size of the particles are determined by Field Emission Scanning Electron Microscopy (FESEM) and software. X-ray diffraction (XRD) is used to confirm the formation of magnetite.

INTRODUCTION

In the last few years many researchers have synthesized several complexes using cyclodextrin (CD). However, most of their works have been focused to the synthesis of CD-organic complexes. A strategy explained by Rauf [1], having a great potential application in materials science, is the modification of CD in order to protect some hydroxyl groups and direct the incoming reagent exclusively to other open hydroxyl groups, this methodology permits the formation of a complex with a variety of guests.

Cyclodextrins are water-soluble oligosaccharides composed of at least six (1-4) linked α-D-glucosyl residues which have the shape of a hollow, truncated cone, capable of forming inclusion complexes with a variety of guest molecules in the solid state, as well as in solution. The size of these molecules should be compatible with the dimensions of the cavity [2]. The inner diameter of the cavity varies from 0.57 to 0.95 nm as the number of glucose units increases from 6 (α–CD) to 8(γ–CD) [3].

Iron oxides have been extensively studied for applications in biomedical areas. Due to its interesting properties and great stability, magnetite is the most popular magnetic material in applications like hyperthermia, drug delivery, cell separation, and others [4-8]. For these applications it is very important to use particles of the iron oxides of a few nanometers in size and narrow particle size distribution in order to improve their magnetic properties.
In the past few years, magnetite nanoparticle has been synthesized by several methodologies, such as sol-gel processing, microemulsion, sputtering, thermal decomposition, and others [9-14].

However the co-precipitation method is still the most popular one due to its simplicity and easy way to control and manipulate the size and morphology of the particles. In the present work, the preparation of magnetite particles using α and β-cyclodextrin is investigated. The methodology thus developed represents an easy route to control the particle size of the magnetite particles by restricting the growth of the particles by the presence of CD. This route can be extended for the synthesis of other inorganic materials as well.

EXPERIMENTAL PROCEDURE

The magnetite-cyclodextrin (M-CD) complex is prepared as follows, 0.086 M ferrous chloride (FeCl$_2$.4H$_2$O) and 0.043 M ferric chloride (FeCl$_3$.6H$_2$O) solutions are mixed, then CD acquired from Aldrich Corporation is added in 0, 0.1, 0.5, 0.75, 1 and 2 wt% concentration in relation to the magnetite. This solution is stirred during 24 hours in order to promote an interaction between CD and Fe ions. A precipitate is obtained after the pH of the solution is adjusted to 11 with ammonium hydroxide. In order to remove the residual ions, the resulting precipitate is centrifuged and washed several times until a pH of 7 is obtained. The precipitate is dried at 100 °C for 2 hours.

The interaction of magnetite with cyclodextrin is determined by FT-IR spectroscopy using a 500 Buck Scientific spectrometer, with spectra of the magnetite and the M-CD complex being recorded in the 400-4000 cm^{-1} range. X-ray diffraction (XRD) is used to confirm the formation of magnetite. Images of the M-CD particles are obtained by using a Field Emission Scanning Electron Microscope (FESEM) with 15KeV and a work distance of 4 -4.3 mm. The samples are dispersed in alcohol by ultrasonic vibration for 5 minutes, then a drop is put on a carbon tape for the nanoparticles to be analyzed. The particle size distribution is determined by means of the Scandium software thought SEM images, the media and standard deviation is obtained by measuring 100 particles of magnetite.

RESULTS AND DISCUSSION

Figure 1 shows the average size of the magnetite particles as a function of the α and β CD content, this plot shows an important reduction in the particle size of the magnetite and its distribution becomes narrower when the complex is formed either with α or β CD. With a 0.5 wt % of αCD the obtained particles have an average diameter of 6 nm that compares to 53 nm without CD. The maximum reduction of particle size is found for a α CD content of 1 wt % to produce a mean size of 5 nm. The major impact is the reduction of the distribution of the particle size, without CD the standard deviation is 15 nm and with 0.5 wt % it is 1.1 nm. In the case of β CD, the maximum reduction of the particle size (around 85%) is found for a content of 0.75 wt % i.e., 8 nm. The use of α and β CD induce the same behavior with a considerable decrease of the average particle size, in this way the particle growth is restricted by the utilization of these oligosaccharides.

The morphology of the magnetite particles in all cases is spherical, as shown in figure 2. In the images it is easy to see that the particles obtained with the use of α–CD (Fig. 2a) and β–CD a particles of few nanometers are obtained.

80

Figure 3 shows the infrared spectra (FT-IR) of Cyclodextrin, magnetite and the M-CD complex. The band at 575 cm^{-1}, in the spectra of magnetite and the complex, is associated with the stretching and torsional vibration modes of the magnetite Fe-O bonds in tetrahedral sites (see arrows in Fig. 3). On the other hand, the band at 3400 cm^{-1} is due to the protonation states by extensive hydrolysis of interfacial water molecules with the magnetite and cyclodextrin (arrows in Fig. 3).

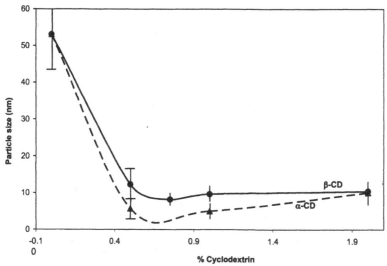

Figure 1. Variation of the mean particle size as a function the α and β CD content.

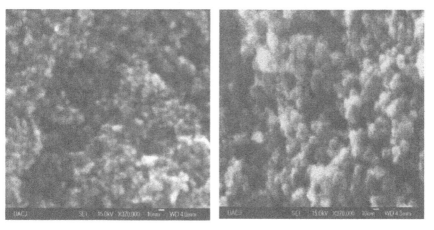

Figure 2. FESEM images of (a) M-0.5 $\alpha\beta$–CD and (b) M-0.5 α–CD complexes.

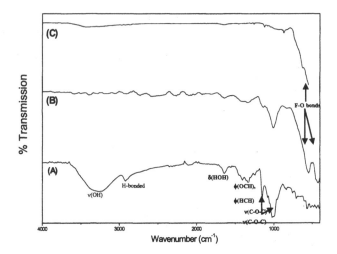

Figure 3. FT-IR of (A) α-Cyclodextrin, (B) magnetite-cyclodextrin complex and (C) Magnetite.

Table 1 presents the interpretation of the FT-IR spectra. In the case of magnetite-cyclodextrin complex, a band for both components is found suggesting that the cyclodextrin structure remains intact. Another important feature in the spectrum is the intensity reduction of the νO-H band at 3400 cm⁻¹ corresponding to the M-CD complex as result of the deprotonation of the hydroxyl groups. This suggests the existence of a bond between magnetite and CD in the complex. On the other hand, in accordance with a previous investigation, the suppression of ¯OH vibrational modes in the 3000 - 3700 cm⁻¹ region can be taken as evidence of a host-guest interaction. All M-CD complexes present the same band as can be seen in Table 1 [1].

Table 1. Assignment of IR bands to Cyclodextrin, M-CD complex and Magnetite.

Wavenumber [cm⁻¹]			Assignment
α-Cyclodextrin	M-CD complex	Magnetite	
3272		3272	ν(OH) of secondary OH groups
2925			H-bonded in different ways
1642	1642		δ(HOH) existing in the cavities
1409	1049		ϕ(OCH), ϕ(HCH)
1328	1328		ϕ(OCH), ϕ(HCH),
1151	1151		ϕ(CCH), ϕ(OCH), ϕ(COH),
1081			ν(C-O-C)
1024	1024		ν(C-O-C)
	549	549	F-O bonds in tetrahedral sites
	422		F-O bonds in octahedral sites

Figure 4 shows the X- ray diffraction pattern of the magnetite, magnetite αCD and magnetite-βCD complexes, all of them have the same crystalline planes that correspond at the spinel structure of the magnetite. At small angles the base line of the magnetite-CD has different behavior, this is attributed to the presence of the amorphous CD.

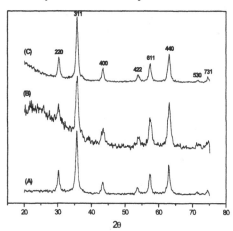

Figure 4. XRD of (A) Magnetite, (B)magnetite-αCD and (C)magnetite β-CD.

CONCLUSIONS

The interaction between the magnetite and cyclodextrin results in the formation of a complex. The average particle size of the magnetite is very similar to the size of the ring of the CD. The growth and the particle size distribution of the inorganic materials are restricted by the use of the cyclodextrin. This approach can be an easy route for the synthesis of inorganic nanoparticles with a few nanometer of size.

ACKNOWLEDGMENTS

The authors are very gratefully with the Science and Technology Council of México for the financial support trough the project CONACYT SEP-2004-C01-47290

REFERENCES

1.   A. Rauf Khan, K. J. Stine, and V. T. Dsouza, Chem. Rev. 98 (1998) 1977-1996
2.   R.D. Sinisterra, M.J Avila-Campo, N.Tortamano, R.G.Rocha,, Inclusion Phenom. Macrocyclic Chem., 40 (2001) 297.
3.   W. Saenger, A. Angew, Chem. Int. Ed. Engl., 19 (1980) 344.
4.   Szehtkum et al, Cyclodextrin thecnology, kluwer academic publisher: Dordrecht, the Netherlands, (1988)
5.   W. L. Sang, B. Seongtae , Y. Takemura, T. M. Kim, J. Kim, H. J. Lee, S. Zurn and C. Sung Kim, J. Magn. Magn. Mats, 310, 2 (2007) 2868-2870.
6.   M.O. Avilés, H. Chen, A. D. Ebner, A. J. Rosengart, M. D. Kaminski and J. A. Ritter, J. Magn Magc Mat, 311, 1 (2007) 306-311.
7.   S. Sieben,C. Bergemann, A. LuKbbe, B. Brockmann,D. Rescheleit Journal of Magnetism and Magnetic Materials 225 (2001) 175-179
8.   Christoph Alexiou,2 Wolfgang Arnold, Roswitha J. Klein, Fritz G. Parak, Peter Hulin, Christian Bergemann, Wolfgang Erhardt, Stefan Wagenpfeil, and Andreas S. Lu¨bbe,CANCER RESEARCH 60, 6641–6648, December 1, 2000]
9.   A Petri-Fink, M. Chastellain, L. Juillerat-Jeanneret. and H. Hofmann, Biomaterials, 26, 15 (2005) 2685-2694.
10.   N.J. Tanga Zhonga., H.Y. Jianga., X.L.Wua, W. Liua., Y.W. Du, J. of Magn and Magn Mats 282 (2004) 92–95.
11.   B.J. Palla, D.O. Shah, P. Garcia Casillas, and J. Matutes-Aquino, Journal of Nanoparticle Research, 2, 1 (1999) 215-221
12.   S. Ohta, A. terada, thin Solid Film, 73(1986) 143.
13.   J. Matutes-Aquino, P. García Casillas, O.Ayala-Valenzuela, and S. García-García, Materials Letters, 38, (1999) 173-177
14.   U. Schewetmann & r.W. Fitzpatrick, Catena Suppl. 21, Catena Verlar., Cremlinge, (1992) 7-30

Mater. Res. Soc. Symp. Proc. Vol. 1243 © 2010 Materials Research Society

Synthesis of Functional Materials by Means of Nitriding in Salts of Al_2O_3–Al, Al_2O_3–Ti and Al_2O_3–Fe Cermets

José G. Miranda-Hernández[1], Elizabeth Refugio-Garcia[1] and Enrique Rocha-Rangel[2]
[1]Departamento de Materiales, Universidad Autónoma Metropolitana
Av. San Pablo No. 180, Col. Reynosa-Tamaulipas, México, D. F., 02200
[2]Departamento de Mecatrónica, Universidad Politécnica de Victoria,
Luis Caballero No. 1200, Col. Del Maestro, 87070, Cd. Victoria, Tamaulipas, México

ABSTRACT

In this work the synthesis of functional gradient materials (FGMs) through the nitriding in salts of previously fabricated cermets is carried out. The matrix for the preparation of the FGMs is made of an Al_2O_3-based cermet with a fine and homogeneous dispersions of metallic particles of Al, Ti and Fe, the final content of the metallic particles in the cermets is of 10 vol. %. These materials are synthesized by powders techniques, considering as variable of study the reinforcement metal in the cermet, previous to sinter, cylindrical samples of the corresponding cermet are uniaxially compacted using 200 MPa. Subsequently, all sintered cermets are characterized by microstructural observations with the help of light microscopy. Obtained cermets are then submitted to a nitriding process by immersion in ammoniac salts. The objective is to attain a compositional gradient of metal-nitrides at the surface. This is done through a chemical reaction between the metallic particles at the cermets' surface and the nitrogen released by the ammonia salts. Functional materials are obtained such as; Al_2O_3–Al-AlN and Al_2O_3–Ti-TiN.

INTRODUCTION

Functional Gradient Materials (FGMs) are materials whose chemical composition depends on the position. This means that they are composite material macroscopically anisotropic. This anisotropy makes the difference between FGM and the traditional cermets. Whereas in a traditional composite material, the composition is constant throughout its volume and its properties are homogeneous to macroscopic scale, the microstructures and properties of a FGM depend on the position in which they are determined [1, 2]. In this type of materials the matrix can be a single phase or composed (several phases). It is well known that alumina-based ceramics possess favorable physical and chemical properties, and additionally a good resistance to corrosion and good thermal stability. Despite these outstanding properties their applications are limited due to brittleness i.e., their deformation leads to cracks after poor ductility [3, 4]. Thus the need for an improved toughness up to high temperatures leads to investigation searching of for new materials. Some investigations have shown that ceramic materials can be strengthened by incorporating small ductile metal particles [4-9]. Thus starting with an alumina based cermet and promoting a compositional gradient from the surface, gives rise to an interesting subject by considering the different structural arrangements that develop and it s effect on the mechanical properties [1].

EXPERIMENTAL PROCEDURE

The experimental route consists of two stages; stage 1 is the synthesis of cermets and stage 2 is the synthesis of FGM. The raw material for stage 1 are powders of Al_2O_3 (99.9 %, 1 μm, Sigma, USA), aluminum, iron and titanium (99.9 % purity, 1-2 μm, Aldrich, USA). The nominal compositions of the final composite materials are Al_2O_3-10 vol. % Al, Al_2O_3-10 vol. % Ti and Al_2O_3-10 vol. % Fe. The original powders are mechanically milled in a horizontal mill at 300 rpm during 12 h and using YSZ balls. The weight ratio balls/powder is of 25:1. The as milled powders are used to produce compacted cylinders (compaction at 200 MPa) with 25 mm in diameter and 3 mm in thickness. These samples are sintered at 1400°C during 1 hour in an argon atmosphere. The heating and of cooling rates are kept constant and equal to 10 °Cmin^{-1}. The characterization of the synthesized products includes the evaluation of density by volume displacement and microhardness measurements with the help of a Vickers indenter. The microstructure of the cermets is investigated by light microscopy (LM). As for stage 2, the samples are subjected to nitriding by thermal treatment in an ammonia salts bath; they are kept at 570 °C for 24 h. Finally, the microstructures of these samples are investigated by scanning electron microscopy (SEM). The SEM is equipped with an energy dispersive X-Rays detector (EDX). Microhardness is also measured in cross sections of these samples (in all cases ten independent measurements per value are carried out).

DISCUSSION AND RESULTS

Cermet fabrication

The relative density of the sintered samples reaches values of 95, 96 and 91 % for the systems Al_2O_3-10 vol. % Al, Al_2O_3-10 vol. % Ti and Al_2O_3-10 vol. % Fe, respectively. Apparently there is a relationship between densification and the type of metal in use. This most likely has to do with the thermal and physical properties of each added metallic component. For example diffusion and distribution of the metal in the final sintered sample, depend on the specific metal that reinforces the ceramic material. The Al containing sample is sintered at a temperature (1400 °C) that promotes the formation of a liquid phase since it is much higher than the melting point of Al. This is done to increase the final relative density of the cermet since the liquid phase can distribute better and join the ceramic particles more efficiently. As for the other materials, the sample with the lowest densification is the composite reinforced with metallic particles of Fe. This is most likely due to the large difference of both thermal expansion and density among Fe and Al_2O_3.

Microstructure

The microstructure as observed by LM for the cermets that contain Al, Ti and Fe is shown in Figure 1. The ceramic alumina matrix can be identified with the opaque and gray phase in Figs. 1a-c, whereas the brighter phase corresponds to the distribution of the metallic particles in the ceramic. In Figure 1a, a homogeneous distribution of aluminum metallic particles can be seen around the ceramic matrix original powder particles, giving as a consequence the good densification previously reported. The typical microstructure of a cermet involves a metallic network surrounding the ceramic material, apparently the synthesized samples have no such network or it is too fine to be resolved by light microscopy. This suggests that wetting in the system Al-Al_2O_3 is too poor. The angle of contact among liquid Al and Al_2O_3 is rather large and higher than 90° [10] and possibly as a consequence Al distribution in the alumina matrix becomes rather uneven. As for Figure 1b corresponding to the composite containing Ti, there is a fine metallic partial network distributed around some grains of the ceramic matrix. This can be achieved by the high specific surface energies reached by titanium during milling inducing a good distribution of this metal in the solid state, since no liquid phases are produced at the sintering temperature. Thus the good densification is most likely related to this phenomenon. Figure 1c shows the microstructure of the composite containing Fe. A good distribution of Fe and the formation of a network are also seen although in this case the average size of the metallic particles is much larger in comparison with the previous samples. This is most likely related to the relatively higher value of the Fe density (7.8 g/cm^3) compared to alumina (3.9 g/cm^3), as well as, the big differences between surface energies of Fe and Al_2O_3. This is expected to lead to Fe segregation towards certain zones of the matrix. In this sample pores are more abundant leading to the decrease in relative densification of this cermet.

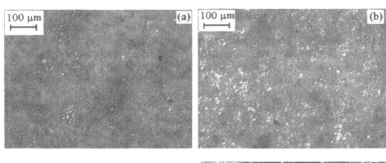

Figure1. Microstructures of cermets after sintering at 1400 °C for 1 h. (a) Al_2O_3-Al, (b) Al_2O_3-Ti, (c) Al_2O_3-Fe.

FGM fabrication

The functional gradient condition in a material is determined by the chemical composition difference across its volume. In this case the chemical gradient is formed by the metallic particles that are localized at the surface of the material. These particles can be nitrided when these cermets are subject to a nitriding process in salts. For this process it starts from the cermets that are made previously in stage 1. The thermal treatment is performed in an ammonia salts bath that is localized inside an electrical furnace. At that temperature, a dissociation of ammonia molecules takes place and Nitrogen is released to diffuse and react with the metal phase at the surface of the composite forming metallic nitrides (AlN, TiN and FeN). The chemical reactions are in Table 1.

Table 1. Nitriding chemical reactions for Al, Ti and Fe.

Chemical reaction	Free energy
$Al + N \rightarrow AlN$	$\Delta G^{\circ}_{f} = -287.0\,\dfrac{KJ}{mol}$
$Ti + N \rightarrow TiN$	$\Delta G^{\circ}_{f} = -308.0\,\dfrac{KJ}{mol}$
$Fe + N \rightarrow FeN$	$\Delta G^{\circ}_{f} = -427.0\,\dfrac{KJ}{mol}$

Microstructure

Figure 2 shows the sample zone near the material surface as observed by SEM. Samples have been cut perpendicularly to the compaction direction to investigate the nitrogen distribution as a function of depth. In SEM images there is a certain surface layer displaying a slightly different color contrast with respect to the Al_2O_3-bulk matrix. Such contrast, in practice exhibited similar texture to the nitride region. So that between the nitride layer and the non-nitride region, there is an intermediate zone which consists of partially-nitride metal particles. Therefore, moving from the outermost surface part into the bulk of material, it has been detected three specific regions, featuring: (1) fully nitride metal particles, (2) partially nitride particles and (3) metallic particles not being nitride. Though for analysis EDX it could not be determine the presence of nitrogen for confirm the formation of metallic nitrides, for this reason is appealed to microhardness measurements in order that indirectly decides the presence of nitrides.

Although the microstructural characterization can be performed for all the synthesized cermets, it is necessary to mention that the composite containing Fe collapsed mechanically after the nitriding process as a result of multiple cracks. As a consequence, the microhardness is not measured in this material. Most likely the synthesis of FGMs based on Al_2O_3-Fe-FeN is not feasible by this method.

Microhardness

The magnitude of microhardness for the different investigated materials is reported in Table 2. In all cases, the measured microhardness is higher for the FGMs in comparison with the corresponding cermet material. It is concluded that nitriding the metallic particle dispersion located at the surface of the ceramic matrix, as conducted in this work, can increase its mechanical strength. The high densification level conferred to the composites is another factor that greatly influences their microhardness value.

The measurements of microhardness are performed as indicated in Figure 3. In the case of Cermets no systematic difference can be noted and the measurements are simply averaged and reported in Table 2. However in the case of the FMG measurements along the cross section show a definite trend. Harder regions are invariably found closer to the surface where the nitriding process has been performed. This is reported in Table 2 as the series of measurements indicated in the FMG columns. There is always a large difference in the microhardness of the surface and bulk for the FMG with an increase around two times the microhardness of the bulk. This confirms the production of a functional material with a composition gradient.

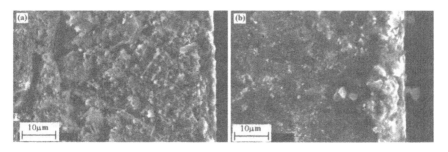

Figure 2. Microstructures of nitrided cermets as observed by SEM. (a) Al₂O₃-Al, (b) Al₂O₃-Ti, (c) Al₂O₃-Fe.

Table 2. Hardness values measured in cermets and FGMs.

System	Composite	FGM	Composite	FGM	Composite
	Al_2O_3-10 vol. % Al	Al_2O_3–Al-AlN	Al_2O_3-10 vol. % Ti	Al_2O_3–Ti-TiN	Al_2O_3-10 vol. % Fe
Hardness (HV)	62.3 +/- 7.1	195.3 +/- 11.3 78.6 +/- 6.2 67.5 +/- 8.1	381.8 +/- 16.4	796.7 +/- 21.7 330.5 +/- 16.8 355 +/- 17.6	112.1 +/- 10.7

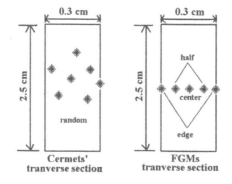

Figure 3. Scheme showing location of microhardness measurement..

CONCLUSIONS

Al_2O_3-based FGMs can effectively be synthesized by inducing fine dispersions of AlN/Al and TiN/Ti, throughout a combination of experimental techniques, such as; mechanical milling, pressureless sintering (argon-atmosphere) and nitriding in ammonia salt process. The later provided that Al_2O_3, Al or Ti and nitrogen precursor monatomic gas are bring together as to react upon sintering forming a functionally-graded-nitrided layer. This *in-situ* synthesis method produces FGMs that are greatly sinterable and do exhibit enhanced microhardness, as compared to Al_2O_3/Al or Al_2O_3Ti cermets. This hardening improvement technique offers the possibility of a low synthesis cost, turning into an attractive synthesis route for scaling the process up to a pilot plant-level.

REFERENCES

1. S. Souto, A. Guitían, R. Francisco and S. De Aza Salvador, Universidad de Santiago de Compostela. Spanish Patent and Trademark Office, Publication number: 2148035, Soliciting number: 009701299.
2. V. Mercedes, Doctoral Thesis, Universidad Autónoma de Madrid, Instituto de Ciencia de Materiales de Madrid, (2003).
3. P. Coca Rebolledo and J. Rosique Jiménez; "Ciencia de Materiales Teoría-Ensayos-Tratamiento", Ed. Piramide, Spain, April 1993.
4. J. Meza: Master Thesis, Universidad Nacional de Medellín Colombia, (2001).
5. J. G. Miranda, Master Thesis Universidad Autonoma Metropolitana, Mexico (2006).
6. J. M. Miranda, S. Moreno, B. Soto and E.Rocha, J. Cer. Proc. Research.7 (2006), 311-314.
7. S. J. Ko, K. H. Min and Y. D. Kim, J. Cer. Proc. Research 3 (2002), 192-194.
8. J. Garcia, W. Lengauer, L. Chen and K. Dreyer, Jornadas SAM 2000 - IV Coloquio Latinoamericano de Fractura y Fatiga, (2000), 1099-1106.
9. A. Feder, l. Llanes and M. Anglada. Bol. Soc. Esp. Ceram. 43, (2004), 47-52.
10. E. Rocha, P. F. Becher and E. Lara, Rev. Soc. Quím., Méx. 48 (2004), 146-150.

Mater. Res. Soc. Symp. Proc. Vol. 1243 © 2010 Materials Research Society

Alumina Extraction from Mexican Fly Ash

Jorge López-Cuevas, David Long-González, Carlos A. Gutiérrez-Chavarría, José L. Rodríguez-Galicia and Martín I. Pech-Canul
CINVESTAV-IPN Unidad Saltillo, Ramos Arizpe, 25900 Coah., México

ABSTRACT

Two alternative chemical methods are studied for the extraction of Al_2O_3 from Mexican Fly Ash (FA). Reaction of FA with H_2SO_4 at high temperature allows extracting ~37% of the total Al_2O_3 contained in the FA as $Al_2(SO_4)_3$, regardless of H_2SO_4 concentration, treatment time and temperature employed. This is partly due to the high chemical resistance of mullite ($Al_6Si_2O_{13}$) contained in the FA. In contrast, reaction of FA with a $CaCO_3$-Na_2CO_3 mixture at 1300°C/1h, followed by lixiviation with a Na_2CO_3 aqueous solution and precipitation of bohemite [AlO(OH)] by addition of either H_2O_2 or NH_4HCO_3, allows extracting ~80% of the total Al_2O_3 contained in the FA as θ-alumina, after calcination of bohemite at 1200°C/1h.

INTRODUCTION

The totality of the alumina (Al_2O_3) consumed by the Mexican ceramic industry is imported, due to a lack of suitable bauxite ($Al_2O_3 \cdot nH_2O$) mineral deposits in this country. Fly Ash (FA) constitutes an attractive alternative raw material for the extraction of alumina. FA is a solid and fine silicoaluminous byproduct of the combustion of powdered coal in power generation plants. It is formed by evaporation of some inorganic components of powdered coal at high temperature. Once these vapors are cooled down, a large number of fine particles is formed with a size ranging from 5 to 75 μm; they can be collected by means of electrostatic precipitators [1]. Chemically, FA is composed by SiO_2, Al_2O_3, FeO + Fe_2O_3, alkali and alkali earth oxides, plus some heavy and transition metal oxides. Mineralogically, FA is composed mainly by mullite ($Al_6Si_2O_{13}$), quartz (SiO_2), magnetite (Fe_3O_4) and/or hematite (α-Fe_2O_3), plus some glassy phases [2]. FA contains about 60-80 wt.% of glassy spheres, the rest of it corresponds to crystalline granules. The spherical particles are constituted by glassy bubbles (cenospheres), solid spheres and cenospheres filled with smaller spheres (plenospheres). The total annual production of Mexican FA is not accurately known, however, in the year 2002 the "José López Portillo" power generation plant, located in the northern state of Coahuila, reported a daily production of ~4,600 tons, from which only ~4-5% is used, mainly by the construction industry. The remaining FA is disposed in landfills. In this work, we present the results obtained from a study carried out on the extraction of Al_2O_3 from Mexican FA, by using acidic and basic chemical methods.

EXPERIMENTAL DETAILS

FA produced by the "José López Portillo" power generation plant has been employed. Iron oxide impurities are separated from the FA by using a high intensity Indiana General dry magnetic separator, with a magnetic field intensity of 5000 Gauss. Two magnetic separation cycles are given to the employed FA batch. The part of the FA that is attracted by the magnetic field due to its high content of paramagnetic minerals, such as Fe_2O_3, is designated as "magnetic

FA fraction", while the part that is repelled by the magnetic field, due to its high content of diamagnetic minerals, such as SiO_2, is designated as "non-magnetic FA fraction". These fractions constitute ~20% and ~80% of the as received FA, respectively. All materials are characterized by laser diffraction (Coulter LS-100), scanning electron microscopy (Philips XL30 ESEM), thermal analysis (Perkin Elmer Pyris Diamond TGA/DTA), X-ray diffractometry (Philips X'Pert, CuKα radiation), and chemical analysis [ICP atomic spectrometry, wet chemistry and X-ray fluorescence (XRF)]. The contents of mullite and glassy phase (plus quartz) are determined by lixiviation with diluted HF (3 wt.%) for 15h [3]. Only the non magnetic FA fraction is employed to carry out the alumina extraction experiments. The acidic method is based on the chemical reaction of FA with concentrated H_2SO_4 at high temperature in order to produce $Al_2(SO_4)_3$. This method has been successfully used for the extraction of aluminum from kaolin [4,5]. In the first step, FA is mixed with concentrated H_2SO_4 (Jalmek, purity of 98%) in a Al_2O_3/H_2SO_4 molar ratio of 1/3. Then the mixture is poured into silica crucibles in order to treat it thermally at 500°C for 3h. Later on, the material is lixiviated with H_2SO_4 1M at 90°C in order to dissolve the $Al_2(SO_4)_3$ formed during thermal treatment. The slow addition of ethylic alcohol (Jalmek, purity of 99.9%) to the resulting solution, under continuous stirring, allows precipitation of the previously lixiviated $Al_2(SO_4)_3$. The basic method consists in a combination of different chemical techniques reported in the literature either for the extraction of alumina and/or silica [6,7] or for the synthesis of Zeolites [8-10] from FA. In this method, FA is first subjected to a silica pre-extraction process, in which it is mixed with an aqueous solution of NaOH 3M (Jalmek, purity of 97%) and treated at 95°C for 3h either in an autoclave or at atmospheric pressure. Then the solid residue is separated by filtration and washed repeatedly with deionized water, until a neutral pH is achieved. This solid residue is dry mixed with $CaCO_3$ and Na_2CO_3 (Jalmek, with purities of 99 and 99.5%, respectively), employing Na_2O/Al_2O_3 and CaO/SiO_2 molar ratios of 1.3/1 and 1.8/1, respectively. After this, the mixture is thermally treated at 1300°C/1h in order to promote the decomposition of mullite, according to the reaction:

$$Al_6Si_2O_{13} + 3Na_2CO_3 + 4CaCO_3 \rightarrow 6NaAlO_2 + 2Ca_2SiO_4 + 7CO_2\uparrow$$

The thermally treated material is lixiviated at 70°C/30 min, employing an aqueous solution of Na_2CO_3 (5%) and a solid/liquid weight ratio of 1/5. This results in the formation of Ca_2SiO_4 solid residue plus $NaAlO_2$ aqueous solution, which are separated by filtration. Later on, the $NaAlO_2$ solution is desilicated by addition of a $Ca(OH)_2$ suspension (Jalmek, minimum purity of 95%). The slow addition of either a 10% H_2O_2 solution (Jalmek, 30% solution) or a NH_4HCO_3 solution (Jalmek, purity of 21.5%) to the desilicated $NaAlO_2$ solution, under continuous and vigorous stirring, until a pH=11.5 is achieved in the first case, or until a pH=10.5 is achieved in the second case, resulted in the precipitation of partially amorphous bohemite [AlO(OH)]. After filtering, washing and drying, this bohemite is thermally treated at 1200°C/1h in order to transform it into alumina.

DISCUSSION

FA characterization

Chemical compositions of as received FA, and magnetic and non magnetic FA fractions, are given in Table I. As can be seen, the as received FA has a $SiO_2+Al_2O_3+Fe_2O_3$ content above 70 wt.%, which indicates that this is a Class F FA, according to the ASTM C618 standard [11]. This kind of FA has a silicoaluminous nature and it comes from fossil coal. This is the most common FA type and its chemical composition is the closest to that of the earth's crust (basalt like) [12]. Table I also shows that magnetic separation decreases the Fe content of FA by ~50%. The ignition losses determined by TGA at 1000°C are assumed to correspond basically to free carbon contained in the materials. The results of HF lixiviation [3] indicates that the FA mullite content is 31 wt.%, with the remaining 69 wt.% corresponding to glassy phase, quartz and other crystalline oxides. The glassy phase contains 11 wt.% Al_2O_3 and 71 wt.% SiO_2.

Table I. Results of chemical analysis by XRF (wt. %). The estimated accuracy is ± 5%.

Oxide	As received FA	Non-magnetic FA fraction	Magnetic FA fraction
*Loss on ignition	2.4	1.9	2.8
SiO_2	60.3	63.7	56.1
Al_2O_3	24.9	25.4	21.2
Fe_2O_3	7.1	4.0	15.6
CaO	2.0	2.1	1.7
Other oxides (TiO_2, MgO, Na_2O, K_2O, etc.)	3.3	3.0	2.7

* Determined by TGA.

SEM analysis indicates that the as received FA is composed mainly by aluminosilicate spheres with a size ranging from 5 to 80 μm (Fig. 1). Some iron oxide particles with a diameter of 10-50 μm are also observed. These oxides are also found incrusted in the aluminosilicate particles. The X-ray diffraction patterns shown in Fig. 2a are similar for all three analyzed materials (i.e., as received FA, magnetic and non magnetic FA fractions). The main effect of magnetic separation is an increased proportion of hematite in the magnetic FA fraction, when compared to the as received FA. The laser diffraction analysis indicates that the mean particle size is reduced by the magnetic separation process, decreasing from 73 μm for the as received FA to 63 μm for the non magnetic FA fraction, which is due to the fact that the mean particle size of the magnetic FA fraction is very large (131 μm). The estimated accuracy for the particle size measurements is ± 5%.

Acidic method used for the extraction of alumina from the FA

The chemical interaction between FA and concentrated H_2SO_4 is very weak at 500 °C, which is evidenced by a slight weight gain after the thermal treatment, without formation of aluminum sulfate. The largest amount of the latter compound is formed when the FA pre-treated at 500°C is heated in a H_2SO_4 1M solution at 90 °C. By using this method, the final alumina extraction is 37% with respect to the total alumina contained in the FA. This low extraction yield cannot be appreciably increased by changing the H_2SO_4 concentration, or by modifying the duration or temperature for the FA treatment. This is partly due to the high chemical stability of mullite, which contains most of the alumina present in the FA. Another likely factor contributing to this is the self-inhibition effect due to the precipitation of calcium sulfate on the surface and within the FA particles, which has been observed by Verbaan and Louw [13] and by Seidel et al. [14, 15], who obtained similar low alumina extractions from FA by using this method.

93

Figure 1. SEM Backscattered Electron Images of: (a) As received FA, (b) Non-magnetic FA fraction, and (c) Magnetic FA fraction.

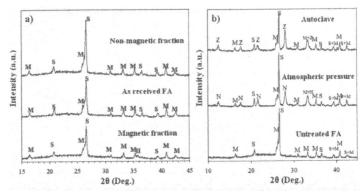

Figure 2. X-ray diffraction patterns for (a) as received FA, magnetic and non-magnetic FA fractions; (b) untreated FA, and FA treated with NaOH either at atmospheric pressure or in an autoclave. S=Quartz, M=Mullite, H=Hematite, N=Na$_{1.4}$Al$_2$Si$_{3.9}$O$_{11.5}$(H$_2$O), and Z=Na$_6$Al$_6$Si$_{10}$O$_{32}$·12(H$_2$O).

Basic method used for the extraction of alumina from the FA

After subjecting the FA to a silica pre-extraction treatment with NaOH, the mean particle diameter decreases from ~64 to 28 μm for treatment in the autoclave, with similar results obtained at atmospheric pressure. Chemical analysis by XRF indicates that only the autoclave treatment results in a significant extraction of silica; the SiO$_2$ content decreases from 63.7 wt.% in the as received FA to 48.7 wt.% in the NaOH-treated FA. The alumina content increases correspondingly from 25.3 to 35.5 wt.%. Figure 2b shows the XRD pattern for FA subjected to the silica pre-extraction treatment. The Na$_{1.4}$Al$_2$Si$_{3.9}$O$_{11.5}$(H$_2$O) phase is detected together with mullite and quartz when this treatment is carried out at 70°C for 3h at atmospheric pressure. Alternatively, a Na$_6$Al$_6$Si$_{10}$O$_{32}$·12(H$_2$O) zeolite is obtained when the treatment is performed at the same temperature and time but in the autoclave. Similar results have been reported in the literature when FA is treated with NaOH in an autoclave [8-10]. The intensity of the silica reflections decreases in the XRD pattern of FA after treatment with NaOH, which indicates that this phase is partially dissolved. The formation of zeolite is not a problem, since the posterior thermal treatment of FA with CaCO$_3$ and Na$_2$CO$_3$ decomposes this phase as well as mullite, giving rise to the formation of NaAlO$_2$, Ca$_2$SiO$_4$ and Na$_2$CaSiO$_4$ (Fig. 3a). When these reaction

94

products are lixiviated with a Na_2CO_3 aqueous solution, a solid residue composed by Na_2CaSiO_4, Ca_2SiO_4 and $CaCO_3$ is obtained (Fig. 3b), together with a sodium aluminate ($NaAlO_2$) solution. These can be separated by filtration. Once the $NaAlO_2$ solution is desilicated by addition of a $Ca(OH)_2$ suspension, and filtrated, partially amorphous bohemite [$AlO(OH)$] is precipitated by addition of either H_2O_2 or NH_4HCO_3 solutions (Fig. 4a). Precipitated bohemite is separated from the remnant solution by filtration. Table II gives the result of chemical analysis of the sodium aluminate solutions, before and after bohemite precipitation.

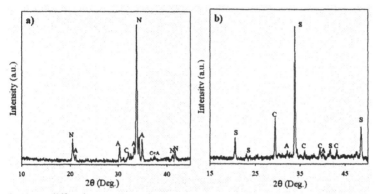

Figure 3. X-ray diffraction patterns of: (a) FA reacted with a mixture of $CaCO_3$ and Na_2CO_3; A=$NaAlO_2$, C=Ca_2SiO_4, and N=Na_2CaSiO_4; (b) solid residue obtained after lixiviation of the reaction products with Na_2CO_3 solution; S=Na_2CaSiO_4, A=Ca_2SiO_4, and C=$CaCO_3$.

Table II. Composition of $NaAlO_2$ and remnant solutions determined by ICP atomic spectrometry. The estimated accuracy is ± 2%.

Solution	Al (g/l)	Si (mg/l)	Fe (mg/l)
$NaAlO_2$ solution	10.4	138.1	<0.042
Remnant solution after desilication and bohemite precipitation	1.8	13.1	<0.042

A mass balance shows that this method allows us extracting ~80% of the total alumina contained in the FA. This alumina yield is lower than that (~88%) reported for similar sintering methods reviewed by V.L. Rayzman et al. [6], which is probably due to the lower sintering temperature used in this work. However, our process allowed us to obtain a final product almost free from Si and Fe impurities (see Table II). Finally, calcination at 1200°C/1h of bohemite produces a fine powder of partially amorphous θ-alumina (Fig. 4b). This is one of the so-called transition aluminas, whose high surface area and reactivity make them suitable for applications in the chemical industry as catalysts; α-alumina, suitable for the ceramic industry, can be easily produced, instead of θ-alumina, by making minor modifications to our final thermal treatment.

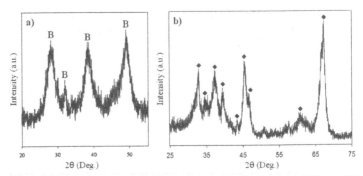

Figure 4. X-ray diffraction patterns of: (a) precipitate obtained after addition of H_2O_2 or $NH_4(HCO_3)$ to the desilicated $NaAlO_2$ solution; B=AlO(OH); (b) θ-alumina ($\blacklozenge$) obtained by heating bohemite at 1200°C/1h.

CONCLUSIONS

A method based on the chemical reaction of FA with concentrated H_2SO_4 at high temperature, allows recovery of ~37% of its total alumina content, regardless of the H_2SO_4 concentration, treatment temperature and time in use. This is partly due to the high chemical stability of mullite, which contains most of the alumina present in the FA. When mullite is decomposed by reacting FA with a mixture of $CaCO_3$ and Na_2CO_3 at high temperature, it is possible to recover ~80% of the alumina contained in the FA. The final alumina product is almost free from Si and Fe impurities.

REFERENCES

1. G. Ferguson, Geotechnical Special Publication No. 36, ASCE, New York, N.Y. (1993).
2. R. Mireles-Álvarez, M. Sc. Thesis, Instituto Tecnológico de Saltillo, Coahuila, México (2004).
3. S. Gomes and M. François, *Cem. Concr. Res.* **30**, 175 (2000).
4. H.K. Kang, S.S. Park, M.M. Son, H.S. Lee, and H.C. Park, *Brit. Ceram. Trans.* **99**, 26 (2000).
5. S.S. Park, E.H. Hwang, B.C. Kim, and H.C. Park, *J. Am. Ceram. Soc.* **83**, 1341 (2000).
6. V.L. Rayzman, S.A. Shcherban, and R.S. Dworkin, *Energy & Fuels* **11**, 761 (1997).
7. R. Padilla, and H.Y. Sohn, *Metall. Trans. B* **16B**, 385 (1985).
8. C.F. Lin, and H.C. His, *Environ. Sci. Technol.* **29**, 1109 (1995).
9. X. Querol, J.C. Umaña, F. Plana, A. Alastuey, and A.L. Soler, *Fuel* **80**, 857 (2001).
10. G.G. Hollman, G. Steenbruggen, and M.J. Jurkovicova, *Fuel* **78**, 1225 (1999).
11. ASTM C618, Annual Book of ASTM Standards, Vol. 04.01, D 3987-85 (1992).
12. L. Barbieri, I. Lancelloti, T. Manfredini, I. Queralt, J.M. Rincon, and M. Romero, *Fuel* **78**, 271 (1999).
13. B. Verbaan, and G.K.E. Louw, *Hydrometallurgy* **21**, 305 (1989).
14. A. Seidel, and Y. Zimmels, *Chem. Eng. Sci.* **53**, 3835 (1998).
15. A. Seidel, A. Sluszny, G. Shelef, and Y. Zimmels, *Chem. Eng. J.* **72**, 195 (1999).

Mater. Res. Soc. Symp. Proc. Vol. 1243 © 2010 Materials Research Society

Rheological characterization and tape casting of aqueous zircon slips

M. León Carriedo, C. A. Gutiérrez Chavarría, J. L. Rodríguez Galicia, M. I. Pech Canul and J. López Cuevas.
Centro de Investigacion y de Estudios Avanzados del IPN. Carretera Saltillo-Monterrey Km. 13,5
C.P. 25900, Ramos Arizpe, Coahuila, México. e-mail: carlos.gutierrez@cinvestav.edu.mx

ABSTRACT

In this work, a commercial zircon flour of 99% purity and mean particle size of 10μm is used. Suspension stabilization process is carried out using the electrostatic stabilization mechanism at pH 11 employing Tetramethylammonium hydroxide (TMAH). Tape casting suspensions are formulated with binding systems obtained by the combination of polyvinyl alcohol with polyethylene glycol, additionally the binding system is added at different binder:plasticizer ratio ranging from 1:1 to 3:1. Suspensions are characterized rheologically determining its type and flow properties by obtaining flow and viscosity curves. Suspensions are tape cast on vinyl-acetate substrates and are dried at room conditions to determine some physical properties of the obtained tapes and to correlate the rheological with the physical properties. The experimental results show that tape casting suspensions have a shear thinning flow type, yield points and flow indexes varying as a function of binder system combination and binder:plasticizer ratio. Also, the binder system has a strong influence on the green tapes characteristics.

INTRODUCTION

Zircon is a material that can be employed in a wide range of industrial and advanced applications at high temperature as structural material and/or in highly corrosive environments. Also, zircon has a low thermal expansion coefficient, which provides high thermal shock resistance and remains structurally stable up to its dissociation temperature at 1676°C. These properties make zircon a very attractive refractory structural material [1-3] and production of zircon tapes would widen its high temperature structural applications. The tape casting process is a method in use to transform suspensions into tapes with large area and low thickness [4]. The suspension of well stabilized ceramic particles is cast onto a flat surface and, after drying, the tapes can be cut in the proper shapes to produce different pieces or multilayer components [5].

A suspension adequate for the tape casting process requires large additions of binders and plasticizers. Binders and plasticizers modify the flow behavior of suspensions since they act as rheological modifiers. Thus they make possible to design the proper rheological flow behavior that requires the processing method [6]. For dry tapes, binders are responsible to ensure good particle cohesion and packing, providing tapes with sufficiently high green mechanical strength to resist handling and storage. On the other hand, plasticizers modify the properties of binders to give them enough flexibility to resist subsequent forming stages and handling.

There are some rheological characteristics that suspensions for tape casting must have to produce good tapes. First, suspensions must have a shear thinning flow behavior to ensure good flow under the blades. Second, they must exhibit a high yield stress to maintain their shape after casting [1]. There are in the literature some rheological flow models that can be used to describe the flow behavior of suspensions. For example, the Heschel-Bulkley model (Eq.1) expresses the rheological behavior (relationship between the flow shear stress and the shear starin rate) of a suspension in terms of its yield strength (τ_0) and flow index (n).

$$\tau = \tau_0 + \eta \cdot \dot{\gamma}^n \qquad (1)$$

In particular, the flow index describes the deviation from a Newtonian flow behavior. The suspension behaves as shear thickening for values of n larger than 1, while n values smaller than 1 represent suspensions with a shear thinning flow behavior [7-9].

EXPERIMENTAL PROCEDURE

The rheological and casting parameters involved in the fabrication of zircon tapes are studied using a commercial zircon (99% purity, American Minerals, USA). The aqueous particle stabilization is carried out at pH 11, employing tetramethyl ammonium hydroxide (TMAH, Aldrich, USA) as a pH controller. A binding system composed by polyvinyl alcohol (PVA, Mw 13000-23000, 98% hydrolyzed, Aldrich, USA) and Polyethylene glycol (PEG, Mw 400, Aldrich, USA) is added to the zircon aqueous suspensions at 10, 15 and 20 wt.%, referred to the solid loading. Additionally, the amount of binder and plasticizer in each binding system is fixed to reach different binder:plasticizer ratios from 1:1 to 3:1 (increasing binder content). Binder is previously dissolved in water and then added to the suspensions to avoid binder aggregates, obtaining 70 wt.% of solid loading in tape to cast slips. Binder dissolution process is carried out following the Bassner method [10], i.e. PVA is dispersed in vigorously stirred water at room temperature and then the solution is heated to 80°C to achieve a complete PVA dissolution.

The rheological behavior of all suspensions is determined by using a rheometer (TA Instruments, Ar2000, UK) capable to operate under controlled shear rate (CR) or controlled stress (CS) conditions. The first mode is used to measure the flow curves and the second for determining the suspensions behavior at low shear rates. The experimental curves are fitted to the Heschel-Bulkley model (Eq. 1) in order to obtain yield stresses and flow indexes of the suspensions. The measurements are performed using a concentric cylinders system, with a rotor diameter of 14 mm at a constant temperature of 25°C. Green density measurements are carried out at room temperature using the Archimedes method and tapes are sintered in air at 1550 °C for 2 h using heating and cooling rates of ±5 °C/min.

RESULTS AND DISCUSSION

Figure 1 shows the flow curve of the primary zircon aqueous suspension without binders. The suspension exhibits a shear thickening behavior with a flow index n=1.64. Shear thickening flow is harmful for all wet ceramic forming processes, including tape casting, for two principal reasons. First, this type of flow implies that viscosity increases with an increasing shear rate. Thus it makes the proper flow of suspensions to reach mold walls or passing through narrow gaps (as found in the tape casting process) very difficult. Second, the shear rate generated during forming is directly related to the green properties of ceramic bodies. This means that the shear rate of the forming process must be high enough to assure flow suspension and a proper particle packing.

The principal characteristics of suspensions for the tape casting process are the existence of a yield stress and a shear thinning flow behavior. Thus the flow behavior of the above described suspension is modified by adding different amounts of the binding component and by varying the binder:plasticizer ratio (B:P Ratio).

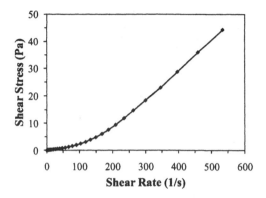

Figure 1. Experimental flow curve of zircon aqueous suspension without binder addition.

Figure 2 shows the flow curves of suspensions after adding 10, 15 and 20 wt% of binding system with different binder:plasticizer ratios. Apparently, all suspensions change their flow type from shear thickening to shear thinning and they all exhibit a well defined yield strength. To validate the flow behavior, the flow curves for the suspensions are fitted to the Herschel-Bulkley model and the flow index and yield strength values are determined. Table 1 shows the corresponding results.

Figure 2a and Table 1 show that viscosity increases with binder content and the flow tends to be more shear thinning for the case of the suspension with 10wt% of binding system. The flow index parameter decreases from 0.96 to 0.89 and the yield stress grows as the B:P ratio increases from 1:1 to 3:1. Figures 2b and 2c show the flow behavior for higher binder additions i.e. 15 and 20 wt%. The resulting suspensions show a more convenient plastic rheological behavior characterized by lower flow indexes. However, as shown in Table I, the yield strength decreases as the B:P ratio increases.

All suspensions are cast on vinyl-acetate substrates. This is done in order to determine their capability to produce tapes with sufficiently high green strength to withstand further processing. The casting parameters employed to produce the tapes are: casting rate of 10 mm/s and blades gap of 350 μm. After casting, all tapes are dried at room temperature under atmospheric conditions during 12 hours. Dried tapes are classified according to its pull out behavior from the substrate and general handling. In the present work, the tapes are further characterized by green density measurements and scanning electron microscopy.

All tapes with a binder content equal to 10 wt% and any B:P ratio and those with 15 wt% and 20 wt% binder content and a 1:1 B:P ratio are very stiff. They exhibit a very poor handling green strength that makes very difficult pulling them out from the substrate without cracking. Apparently, the poor properties of tapes with a binder content of 10wt% can be attributed to deficient particle coverage as a consequence of insufficient binder content. On the other hand, tapes obtained from suspensions with a binder content of 15 wt % and 20 wt % and a 1:1 B:P ratio are stiff as a consequence of insufficient plasticizer to enhance the plasticity of binder. In the case of suspensions prepared with a binder content equal to 15 wt % and B:P ratios of 2:1 and 3:1 and also 20 wt % with a B:P of 2:1, the tapes show good pull out properties and enough green strength to resist further processing. In these cases it is possible to produce one meter long tapes without cracking.

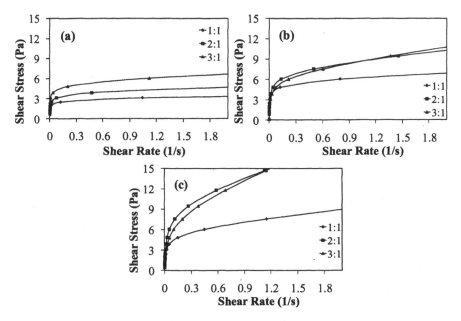

Figure 2. Controlled stress flow curves of suspensions containing (a) 10w%, (b) 15w% and (c) 20w% of binding system at different B:P ratios.

Figure 3, shows the surface microstructure of tapes prepared with the best suspension formulations. As can be seen, the surface of the tapes exhibits good packing, low porosity and good homogeneity. The green density of tapes prepared from suspensions with 15wt.% binding system and B.P. ratios of 2:1 and 3:1 are 59.3 %T (± 4%) and 58 %T (±3%), respectively. The green density of the tape prepared from the suspension with 20 wt% binding system and a B.P. ratio of 2:1 is 55.8 %T (±2%). Although these green density values are rather low, the resulting density of fired tapes is ~ 92 %T.

Table I. Rheological parameters of experimental suspensions (Herschel-Bulkley model)

% B. S.	B:P Ratio	Yield Stress (Pa)	Flow Index (n)
10	1:1	1.96	0.96
10	2:1	2.17	0.95
10	3:1	2.99	0.89
15	1:1	3.09	0.86
15	2:1	3.02	0.77
15	3:1	2.08	0.75
20	1:1	1.93	0.76
20	2:1	1.63	0.56
20	3:1	1.29	0.62

Figure 3. Surface microstructure of tapes prepared from the best suspension formulations: (a) 15 wt% binding system with a B:P ratio of 2:1, (b) 15 wt% binding system with a B:P of 3:1, c) 20 wt% binding system with a B:P ratio of 2:1.

CONCLUSIONS

Zircon suspensions without binders show shear thickening flow behavior. However, it is possible to change this type of flow behavior to shear thinning by adding 10 wt%, 15 wt % and

20 wt % of binder, making them appropriate for tape casting. In addition, the B:P ratio is also important to produce sufficiently high mechanical green properties to resist handling and the pull out operations. From this, the best formulations to produce one piece (about 1 m long) tapes is a combination of relatively high binder concentration (> 10 wt%) and a proper B:P ratio. This ensues in good binder particle coverage and good tape flexibility, respectively. Thus, 15 wt% and 20 wt % at B:P ratios of 2:1 and 3:1 are the best suspensions to produce zircon tapes based on an aqueous system composed by polyvinyl alcohol-polyethylene glycol.

ACKNOWLEDGEMENTS

This work has been supported by CONACyT México under the contract 46720-Y and M. León Carriedo gratefully acknowledges a CONACyT scholarship.

REFERENCES

1. R. Moreno, J. S. Moya, J. Requena, *Bol. Soc. Esp. Ceram. Vidr*, **16**, 165-71 (1990).
2. M. Hamidouche, N. Bouaouadja, R. Torrecillas, G. Fantozzi, *Ceramics International*, **33**, 655-662 (2007).
3. Liliana B. Garrido, Esteban F. Agloetti, *Ceramics International*, **27**, 491-499 (2001).
4. R. Moreno, J. Requena, *Bol. Soc. Esp. Ceram. Vidr*, **31**, 99-108 (1992).
5. Andrea Roosen, "Basic requerements for tape casting of ceramic powders", *The American Ceramic Society*, **1**, 675-692 (1988).
6. Rodrigo Moreno, "The role of slip additives in tape-casting technology: Part I- Solvents and disperants", *Am. Ceram. Soc. Bull.*, **71**, 1521-31 (1992).
7. Bernd Bitterlich, Christiane Lutz, Andrea Roosen, *Ceramics International*, **28**, 675-683 (2002)
8. R. Moreno, "Reología de Suspensions Cerámicas", Consejo Superior de Investigaciones Cientificas, Madrid, Spain (2005) pp. 41,42,44.
9. C. Gutiérrez, A. Javier Sánchez-Herencia, R. Moreno, *Bol. Soc. Esp. Ceram. Vidr*, **39**, 105-117 (2000).
10. S. L. Bassner, E. H. Klingenberg, *Am. Ceram. Soc. Bull.*, **77**, 71-75 (1998).

Mater. Res. Soc. Symp. Proc. Vol. 1243 © 2010 Materials Research Society

Thermoelectric Power Changes of Low Strength Steel Induced by Hydrogen Embrittlement Tests

N. Mohamed-Noriega[1], E. López Cuéllar[1] and A. Martinez de la Cruz[1]
[1]FIME, Universidad Autónoma de Nuevo León, San Nicolás de los Garza, México.

ABSTRACT

This work reports the thermoelectric characterization of a hydrogen embrittlement (HE) of low strength steel. Two sets of tests are performed in an electrochemical cell of H_2SO_4, with and without applied stress, lasting from 2 to 94 hours. Thermoelectric power (TEP) measurements are matched with ductility measurements (%RA and %EL) of samples tested in tension, as well as with microhardness measurements. Results indicate that TEP is sensitive to HE of low strength steels; the maximum variation of TEP is of ~80nV/°C for samples tested without stress.

INTRODUCTION

TEP is a material property which relates electrical and thermal properties. It is the sum of two components: the diffusive component and the lattice component. The diffusive component (S_d) represents the electron contribution and the lattice component or phonon drag (S_g) represents the interactions between electrons, phonons and lattice imperfections. The diffusive component is a linear function of temperature that depends on the evolution of the electrical resistivity (ρ); see equation 1. The lattice component depends on the probability that a phonon bounces with a lattice imperfection (P_i) or with an electron (P_e), see equation 2.

$$S_d = A[\rho(E)] \ T \qquad (1)$$

$$S_g \propto \frac{P_e}{\sum P_i} \qquad (2)$$

For some materials the total TEP can be obtained simply by adding both components ($TEP = S_d + S_g$). However, for iron and its alloys (e.g. steel) the explanation of the total TEP by the two components has not been completely satisfactory and the deconvolution of the two components is uncertain [1].

From the definition of TEP, it can be deduced that any modification on the composition, the crystal structure or the microstructure of the material will alter its value. Kawaguchi et al. studied successfully the evolution of TEP as a function of Cr content in a cast duplex stainless steel [2]. López Cuéllar et al. have used TEP to analyze the growth of the different phases in an Inconel 718 superalloy [3]. Caballero et al. explored the variation in the TEP as a function of martensite content by quenching a stainless steel at different temperatures [4].

Steel ASTM A-572 grade 50 is frequently used in the manufacture of pipelines. In such a case, some of the hydrogen in the hydrocarbons flowing trough the pipelines is absorbed into the metal, producing hydrogen embrittlement. In the present study, TEP is used to follow the evolution of the ductility of the material as it is charged with hydrogen. In order to investigate the variation of ductility and to correlate it with the TEP, Vickers microhardness, %RA and %EL are determined.

EXPERIMENTAL PROCEDURE

A commercial steel, ASTM A-572 grade 50 (API 5L), is used for hydrogen embrittlement tests. Hydrogen charging is carried out electrochemically on samples with and without applied stress in a H2SO4 solution with different molarities (see table I) and lasting from 2 to 94 hours. The molarity variation of the solutions responds to the objective of avoiding observable corrosion in the samples; corrosion that may affect the TEP measurements.

Table I. Molarities of the H_2SO_4 solutions according to the charging time and the stress condition

	Molarity	
	Without Stress	**With Stress**
0 – 30 hours	1.0 M	0.6 M
30 – 60 hours	1.0 M	0.4 M
60 – 94 hours	1.0 M	0.2 M

The electrochemical cell consists of a gaseous hydrogen anode and the sample as cathode. A current density of 33-66 mA/cm^2 and a hydrogen flow of 260±60 mL/min are used, as recommended by other authors [5-6]. The samples have plate geometry (5cm x 1.5cm x 0.2cm) and are polished for both sides up to a 4000 grinding paper, prior to hydrogen charging. The samples with applied stress are subjected to an *in situ* bending stress by a mechanical device, placed inside the electrochemical reactor. It produces a deflection of ~0.5 mm.

Prior to hydrogen charging, the samples are characterized by measuring the TEP and Vickers microhardness. After the test, the samples are characterized again by measuring the TEP, Vickers microhardness and determining the %RA and %EL from tensile tests.

The TEP has been measured in a TechLab equipment with two steel blocks. The blocks are separated and kept at two different temperatures (15°C and 25°C). The samples are fixed with pressure-screws above the blocks and the voltage produced across the samples is measured as a function of temperature (see Figure 1); TEP measurements are done in four directions, two upside and two upside-down, with the flat sides perpendicular to the pressure-screws.

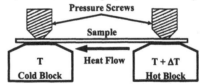

Figure 1. Diagram of the TechLab equipment with the sample in place

Vickers microhardness is measured in a Shimadzu 341-64278 Type M equipment. Measurements are carried out in six points across the longitudinal axis of the sample (three on each side) using a 500 gr load and a 15 seconds indentation time.

Tensile tests are performed in an Instron 8502 tensile machine at a cross head speed of $1^{mm}/_{min}$ without extensometer. The %RA and %EL are calculated considering the cross section area and length of the samples before the test and after fracture; these two variables are widely used as embrittlement parameters [7-8]. These calculations are done with the purpose of estimating the degree of embrittlement of the material and relate it with the TEP measurements.

Additionally, a set of interrupted tensile test (see table II) have been performed on samples of the same material. This done to determine the isolated effect of plastic strain on the TEP and separate it from the global effect due to the HE. A MTS type 793 tensile machine is used with a crosshead speed of 1 mm/min without extensometer. The TEP of the samples is measured before and after the tensile test.

Table II. Interrupted tensile test conditions

Sample	Strain (%)
1	10%
2	20%
3	40%
4	60%
5	80%

RESULTS

The stress-strain curves obtained from interrupted tensile tests are shown in figure 2a. Figure 2b shows the variation of ΔTEP as a function of the strain, where ΔTEP is the difference between the value after (TEP_{After}) and before (TEP_{Before}) the interrupted tensile tests; ΔTEP decreases as a function of deformation. This behavior has been previously observed in TiNi [9] and pure iron [10].

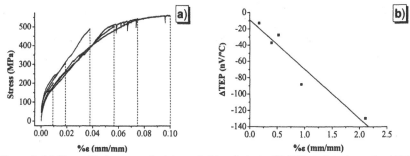

Figure 2. (a) Stress-Strain curves for interrupted tensile tests. (b) Effect of strain on the ΔTEP ($\Delta TEP = TEP_{After} - TEP_{Before}$).

The impact of hydrogen charging on the TEP is shown in figure 3. Figure 3a shows the variation of ΔTEP as a function of the charging time for samples with stress, while figure 3b shows the results for samples without stress; ΔTEP is the difference between the value after (TEP_{After}) and before (TEP_{Before}) the hydrogen charging. As an aid to the reader, a quadratic fitting curve is calculated and is drawn in both graphs. Both curves seem to exhibit a critical point at around 40 hours of charging (maximum for samples without stress and minimum for samples with stress), instead of presenting a linear relationship as in figure 2b. This behavior can be explained on the basis of two different and opposite effects that modify simultaneously the TEP. A first effect can be deduced from figure 2b, where a decrease of the TEP can be associated to the distortion of the lattice; as suggested for pure iron [11]. The second effect is an increase of

105

the *TEP* associated to an increase in the solute concentration [10]. The fact that the samples with stress (see figure 3a) display an opposite behavior to the samples without stress (see figure 3b), can be explained by the initial distortions of the lattice as a result of the applied stress.

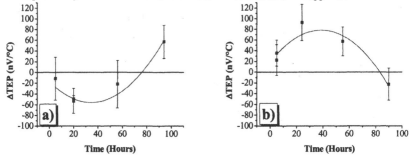

Figure 3. Variation of the *TEP* ($\Delta TEP = TEP_{After} - TEP_{Before}$) as a function of the charging time for the samples (a) with stress and (b) without stress.

Figure 4a and 4b show the variation of Vickers microhardness (ΔVH) as a function of the charging time in samples with and without applied stress respectively. ΔVH is the difference, between the value after (VH_{After}) and before (VH_{Before}) the hydrogen charging. Just as for the graphs in figure 3, a quadratic fitting curve is calculated and is drawn in both graphs. Curve 4a seems to exhibit a maximum approximately after 50 hours of charging, while curve 4b after 65 hours. The behavior of the samples with stress is drastically reversed after the maximum, while for the samples without stress the reversing in the behavior is not that evident.

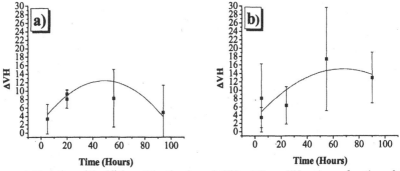

Figure 4. Variation of the Vickers Microhardness ($\Delta VH = VH_{After} - VH_{Before}$) as a function of the charging time for the samples (a) with stress and (b) without stress

The *%RA* and *%EL* calculated from the cross sections and lengths of the samples tested under tension are shown in figure 5, as a function of charging time for the samples with stress, while figure 5c and 5d show the results for samples without stress. Each point of the four plots from figure 5 is a single measurement; therefore, no error bar can be offered. As above, a quadratic fitting curve is calculated and drawn in the four graphs. The justification of the use of a quadratic fitting for the data of figure 5 is not as evident as for figure 3 and 4 but, since figure 5 represents a different parameter for the same phenomena, then a similar behavior can be expected.

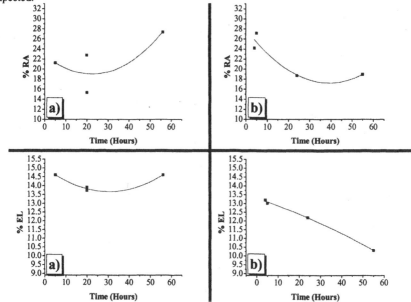

Figure 5. Variation of the *%RA* and the *%EL* as a function of the charging time for the samples with stresses (a-b) and without stress (c-d).

DISCUSSION

The results of Vickers microhardness (Fig. 4), %RA (Fig. 5) and %EL (Fig. 5) of the samples charged with and without stress show a good consistency among themselves and a behavior similar to that of the TEP (Fig. 3). This behavior can be explained as a result of two opposite effects: an embrittlement at short charging times and a softening at longer times. These apparently contrasting effects on the mechanical properties of materials have been previously reported by several research groups for different materials, including steels [12-14]. The embrittlement is often associated to the accumulation of hydrogen in specific regions of the microstructure, reducing the cohesive strength of the lattice, promoting dislocation generation and eventually nucleation and growth of cracks. In contrast, the softening is attributed to an increase in the dislocation mobility. The aim of this work is not to explain the sources of this contrasting behavior, but to assess if TEP can describe it.

On the other hand, in order to understand the effect of hydrogen charging (with and without stress) on TEP, two general effects must be consider, they both take place during charging. First, the increase in hydrogen concentration as a result of charging time and second, the presence and increase of the lattice deformation produced by both hydrogen and the externally introduced stress. As above discussed, TEP is explained as the sum of two components, the diffusive component (S_d) and lattice component (S_g). Thus its observed behavior can be explained by analyzing the variation of S_d and S_g. However the tests have been carried out at room temperature and small deformations are externally produced in the material, the lattice component of the TEP can be neglected, thus the behavior of the TEP can be explained only as a function of the diffusive component. S_d depends only on the electrical resistivity (see equation 1). This parameter is given by $\rho = (m_e^*/n \tau e^2)$ where m_e^* is the effective mass of the electrons, e the electron charge, n the concentration of charge carriers and the average time between collisions (τ) of electrons with phonon and/or lattice imperfections. This shows that ρ, and thus S_d, is only a function of n and τ. The charge carrier concentration (n) is directly related to the hydrogen concentration, which according to previous work increases with charging time [5, 14]. Now, assuming that each absorbed hydrogen atom adds to the charge carrier concentration, a decrease in the diffusive component of the TEP can be expected and this becomes an increase in the ΔTEP value. Moreover, the presence of hydrogen atoms in the crystal structure of steel promotes lattice distortions, which are also known to reduce the ΔTEP value (Figure 2b). This reduction can be understood as a consequence of a shorter average time between collisions (τ), promoted by the increase in the density of lattice distortions, which manifest as an increase of S_d, too. Thus in spite of the possible detailed explanations for the contrasting behavior of the process, the TEP shows a similar behavior as the mechanical properties, implying its capacity to follow the evolution of hydrogen embrittlement phenomena.

CONCLUSIONS

The results presented in this work, show that the TEP is sensitive to the processes involved in hydrogen charging. The TEP is controlled by two opposite effects: the increase in ΔTEP produced by the rise of the hydrogen charging time and the decrease of ΔTEP due to lattice distortions. However, lattice distortions show a stronger influence on the TEP than hydrogen concentration. Finally, TEP is a practical technique to characterize the percentage of deformation in a steel ASTM A-572 grade 50.

REFERENCES

1. J. M. Leborgne, PhD Thesis (No. 96ISLA0134), INSA-Lyon, 1996.
2. Y. Kawaguchi y S. Yamanaka, J. of Alloys and Compounds, Vol. 336, p. 301-314, 2002.
3. E. López Cuéllar et al., J. of Alloys and Compounds, Vol. 467, p. 572-577, 2009.
4. F.G. Caballero et al., Scripta Materialia, Vol. 50, p. 1061-1066, 2004.
5. G. Benamati et al., J. of Nuclear Materials Vol. 212-215, p. 1401-1405, 1994.
6. M. Gojic, L. Kosecb and P. Matkovic, Eng. Failure Analysis Vol. 10, p. 93-102, 2003.
7. P. Lacombe et al., HE and SCC, ASM, pp. 79-102, 1984.
8. M. Bernstein et al., HE and Stress Corrosion Cracking, ASM, pp. 135-152, 1984.
9. E. Lopez Cuellar, G. Guenin and M. Morin, Mat. Sci. Eng. Vol. A358, p. 350-355, 2003.
10. R. Borrelly and D. Benkirat, Acta Metall Vol. 33 No. 5, p. 855-866, 1985.
11. D. Benkirat, P. Merle, and R. Borrelly, Acta Metall Vol. 36 No. 3, p. 613-620, 1988.
12. F.A. Lewis, Pure & Appl. Chem., Vol. 62, Issue 11, p. 2091-2096, 1990.
13. I. M. Robertson, Engineering Fracture Mechanics, Vol. 68, p. 671-692, 2001.
14. F. W. H. Dean and S. W. Powell, NACE International, No. 04472, pp. 1-14, 2004.

Mater. Res. Soc. Symp. Proc. Vol. 1243 © 2010 Materials Research Society

Effect of Hot Band Annealing on Microstructure of Semi-Processed Non-Oriented Low Carbon Electrical Steels

Emmanuel J. Gutiérrez Castañeda and Armando Salinas Rodriguez
Centro de Investigación y de Estudios Avanzados del Instituto Politécnico Nacional, Saltillo Coahuila, P.O Box 663, México 25900.

ABSTRACT

Effects of hot band annealing on the final microstructure and magnetic properties of cold rolled and annealed non-oriented grain Si-Al electrical steel strips are investigated. Microstructures are characterized using optical and scanning electron microscopy and magnetic properties are determined using a vibrating sample magnetometer. It is shown that annealing of hot rolled bands at temperatures between 800 and 850 °C causes rapid decarburization and development of a microstructure consisting of large columnar ferrite grains free of secondary particles. This microstructure leads, after cold rolling and a fast annealing treatment, to large grain microstructures similar to those observed in production scale, fully processed strips. It is observed that the final grain size increases with the final annealing temperature, leading to a significant improvement of the magnetic properties. Therefore, hot band annealing technology can be an attractive alternative processing route for the manufacture of non-oriented grain low carbon Si-Al processed electrical steel strips.

INTRODUCTION

Electrical steel is a soft magnetic material used for cores of electrical generators, transformers and motors [1]. For these applications, low core loss and high permeabiity are required [2]. Magnetic properties are mainly dependent on two microstructural parameters: grain size and texture [3]. Therefore, to achieve low core loss in materials that are used after stress relief annealing, grain growth is necessary [2]. Some processing routes have been reported for production of high permeability electrical steel grades, such as optimized hot rolling schedule [4], hot-band annealing [5] and cold rolling in two steps with intermediate annealing [6]. Some authors have studied the effect of hot-band grain size on magnetic properties of non-oriented electrical steels [7]. They found that hysteresis losses decreased with increasing hot-band grain size. According to this, hot-band annealing could change the grain structure by enhancing grain growth, which can help to improve magnetic properties of these materials. In the present investigation, the effect of hot band annealing on microstructure of semi-processed non-oriented low C electrical steels has been investigated.

EXPERIMENTAL PROCEDURE

Hot rolled bands, 2.7 mm in thickness, of commercial grain non-oriented (GNO) low C electrical steel are obtained from a local steelmaker. Table I lists the chemical composition of the hot band as determined by optical emission spectrometry. This material is produced by hot rolling of 50 mm thick continuous cast thin slabs. Rectangular, 30 mm wide by 60 mm long samples are cut from the hot band. The long direction of the samples is parallel to the original hot rolling direction. The samples are heated at rate of 15 °C/min and annealed in dry air at temperatures of 700, 800, 850, 900, 950 and 1050 °C in a muffle-type furnace. Annealing times are 10, 20, 60 and 150 minutes and cooling is performed in air. A sample of the as-received hot-rolled band and the annealed hot-rolled band samples are cold-rolled to a final thickness of 0.57 mm using a two-high laboratory rolling mill. Finally, the cold-rolled steel sheets are annealed during 3 minutes at 700, 850 and 1000 °C. The microstructures of longitudinal sections of the resulting samples are characterized using optical and scanning electron microscopy. Finally, the magnetic properties are measured at 1.2 T and 60 Hz in a Lakeshore 7000 VSM magnetometer.

Table I. Chemical composition of steel [wt %]

C	Si	Al	S	Mn	P	Mo	Cu	Cr	Ni
0.05	0.57	0.21	0.004	0.32	0.042	0.011	0.029	0.019	0.029

RESULTS AND DISCUSSION

Effects of hot band annealing on the microstructure

In general, hot band annealing treatments during up to 25 minutes did not cause a significant effect on grain size at any temperature. However, longer annealing times produced important microstructural changes. Figure 1 illustrates the effect of annealing temperature on the microstructures of the hot-rolled electrical steel specimens treated during 150 minutes. As can be seen, the effects are rather complex. Annealing at temperatures lower than 800 °C causes only a marginal effect on the grain size, as compared with the as-received condition (compare Figs. 1a and b). Samples annealed at T ≥ 900 °C exhibited uniform microstructures of equiaxed ferrite grains with sizes larger than those observed in the "as received" material or in hot band samples annealed at T ≤ 750 °C (see Fig. 1d). In contrast, samples annealed at 800 and 850 °C exhibited a peculiar microstructure consisting of large columnar ferrite grains (Fig. 1c). This type of microstructure is observed in samples annealed at 850 °C even for annealing times shorter than 60 minutes. According to published Fe-Si phase diagrams for carbon contents between 0.05 and 0.08 [8], the Ae_1 and Ae_3 transformation temperatures for a Fe-0.6Si-0.2Al alloy (similar to the composition of the present steel) are around 730 y 880 °C, respectively. Therefore, the annealing temperatures at which large columnar grains develop must be high in the ($\alpha+\gamma$) phase field. Although the origin of columnar grains is not investigated further in the present work, it appears that fast decarburization in the ($\alpha+\gamma$) intercritical region favours this type of columnar growth.

Fig. 2 shows that hot band annealing at 850 °C causes decarburization of the strip in about 60 minutes. In contrast, decarburization at 700 and 1050 °C did not occur. Columnar grain growth in non-oriented electrical steels has been reported by other researchers [9]. Decarburizing annealing in the two-phase region is known to produce columnar grain growth and improve magnetic properties. Methods to produce large columnar grains in these materials include long-term annealing in vacuum and subsequent decarburizing annealing in a wet hydrogen atmosphere. Kováč, Džubinský and Sidor [9], proposed a "two-step" decarburizing continuous annealing process in the intercritical region to produce columnar grain microstructures in the final sheet.

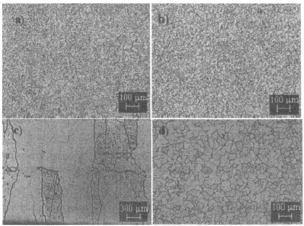

Figure 1. Effect of annealing temperature on the microstructure of hot-rolled GNO low C steel specimens annealed during 150 min. (a) Hot-rolled condition (b) 700 °C, (c) 850 °C, (d) 1050°C.

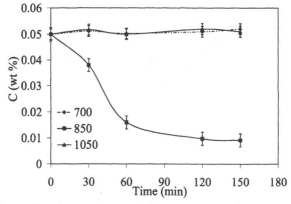

Figure 2. Effect of annealing time on the overall C content of hot band samples annealed at 700, 850 and 1050 °C.

The second phase distribution also changed significantly depending on the annealing temperature. The as-received hot band exhibited small regions of pearlite (α-Fe+Fe$_3$C) at ferrite grain boundary triple points and some isolated, fine particles of Fe$_3$C formed mainly at ferrite grain boundaries (Fig. 3a). Samples annealed at 700 and 850 °C (i.e T$\leq$Ae$_3$) during 150 minutes did not show the presence of pearlite or isolated Fe$_3$C particles (Fig. 3b). The absence of pearlite and Fe$_3$C particles is due to second phase particle dissolution and decarburization during annealing at temperatures below Ae$_3$. In contrast, mcrostructures of samples annealed at T > Ae$_3$ in the austenite phase field (Fig. 3c) showed the presence of pearlite islands. This observation indicates that decarburization in the present electrical steel is much faster when ferrite and austenite are both present in the microstructure.

Figure 3. Secondary phases in: (a) Hot-rolled as-received material, (b) sample annealed during 150 minutes at 850 °C and (c) sample annealed during 150 minutes at 1050 °C.

Effect of hot band annealing on the final microstructure and magnetic properties.

Cold rolling (79% reduction in thickness) of the as-received hot band and annealed samples produced microstructures of severely deformed grains elongated parallel to the rolling direction. However, the length and thickness of the deformed grains depended on the size of the original grains prior to deformation.

Figure 4 shows the effect of temperature of the final 3-minute annealing treatment on the microstructure of samples subjected to hot band annealing at various temperatures. As can be seen, hot band annealing at 700 and 1050 °C applied prior to cold rolling and final annealing resulted in fully recrystallized microstructures with very fine and uniform grain size (Figs. 4a, 4c, 4d and 4f). In contrast, when the hot band annealing temperature is increased to 850 °C, fully recrystallized microstructures are only observed when the final annealing temperature is increased to 850 °C (Figs. 4e). Evidently, the effect of final annealing on the microstructure depends strongly on the characteristics of the previous microstructure produced by the hot band annealing treatment. These observations have important practical implications. Electrical steels with good magnetic properties (low coercivity, high permeability and low hysteresis losses) require large grain sizes [6, 7]. In the manufacture of semi-processed electrical steels, the final heat treatment is optimized so that relief of internal residual stresses, decarburization and grain growth all contribute to achieve the highest possible magnetic quality.

The results of the present work show that decarburization annealing in air at temperatures between 800 and 850 °C of the present hot rolled steel strip produced very large columnar grains. After cold rolling and final annealing during 3 minutes at temperatures higher than 800 °C, a fully recrystallized grain microstructure is obtained. Fig. 5 shows the microstructures of the samples subjected to hot band annealing during 150 minutes at 850 °C and final annealing during 3 minutes at 850 (Fig. 5c) and 1000 °C (Fig. 5d). In the same figure are included the microstructures of production semi-processed (Fig. 5a) and fully processed (Fig. 5b) strips of the same steel. As can be seen, increasing the final annealing temperature increases the final grain size of the material subjected to the hot band anneal. Moreover, when the final anneal is performed at 1000 °C, the grain microstructure of the material is nearly identical to that observed in the production fully processed strip.

Figure 4. Effect of hot band annealing temperature on the final microstructures after cold rolling and a 3 minute annealing treatment (HBA=temperature of hot band annealing, FA=temperature of final annealing).

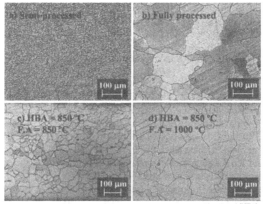

Figure 5. Microstructure of (a) semi-processed and (b) fully processed GNO low C electrical steel produced industrially. Micostructure of samples subjected to hot band annealing during 150 minutes at 850 °C, cold rolled and finally annealed during 3 minutes at (c) 850 °C and (d) 1000 °C.

Table II shows a comparison of the magnetic properties of the production strips with those of the hot band samples annealed during 150 minutes at 850 °C. As can be seen, as the temperature of the final annealing is increased there is a significant improvement of the magnetic properties with respect to those obtained in the production semi- or fully- processed strips. These results show that hot band annealing prior to cold rolling and final annealing is an attractive alternative processing route for the manufacture of fully processed grain-non-oriented Si-Al electrical steel strips for the construction of magnetic cores for electrical motors.

113

Table II. Effect of hot band annealing (150 minutes at 850 °C) and final annealing temperature on the magnetic properties of grain non-oriented Si-Al electrical steel strips

	Coercivity (Oe)	Remanence (G)	Energy losses (W/kg)
Industrially semi-processed strip	38.78	583	3.47
Industrially fully processed strip	37.58	597	3.19
Annealed at 700 °C	24.34	430	2.14
Annealed at 850 °C	25.09	395	1.54
Annealed at 1000 °C	24.27	386	1.90

CONCLUSIONS

Hot band annealing technology appears to be an attractive alternative processing route for the manufacture of grain non-oriented low C, Si-Al processed electrical steel strips. The results of the present work show that annealing of hot rolled bands at temperatures between 800 and 850 °C causes rapid decarburization and development of large columnar ferrite grains free of secondary particles. This microstructure results, after cold rolling and a fast annealing treatment, in microstructures very similar to those observed in production-type fully processed strips but with superior magnetic properties (smaller coercivity, remanence and hysteresis losses).

ACKNOWLEDGMENTS

The financial support from CONACYT is duly recognized.

REFERENCES

1. Takeshi Kubota, Masato Mizokami, Masahiro Fujikura, Yoshiyuki Ushigami, Electrical Steel Sheet for Eco-Design of Electrical Equipment, Nippon Steel, 81, 53 (2000).
2. Atsuhito Honda, Yoshio Obata, Susumu Okamura, History and Recent Development of Non-Oriented Electrical Steel, Kaisaki steel, 39, 13 (1998).
3. Jong Tae Park and Jae Kwan Kim, Recrystallization, Grain growth and Texture Evolution in Non-Oriented Electrical Steels, Posco Steel 10, 26 (2007).
4. O. Fischer and J. Schneider, Influence of Deformation Process on the Improvement of Non-oriented Electrical Steel, J. Magn. and Magn. Mater, 254, 302 (2003).
5. Rubens Takanohashi and Fernando Jose Gomes Landgraf, Effect of hot-band grain size and intermediate annealing on magnetic properties and texture of non-oriented silicon steels, J. Magn. and Magn. Mater. 304, 608 (2006).
6. Sebastião C. Paolinellia and Marco A. da Cunha, Development of a new generation of high permeability non-oriented silicon steels, J. Magn. and Magn. Mater, 304, 596 (2006).
7. Chun-Kan Hou, Effect of Hot Band Annealing Temperature on the Magnetic Properties of Low-carbon Electrical Steels, ISIJ International, 36, 563 (1996).
8. Harold E. McGannon, *The Making Shaping and Treating of Steel*, ninth edition (United States Steel Corporation), Pittsburgh, Pennsylvania, 1161, (1970).
9. F.Kováč, M. Džubinský and Y.Sidor, J. Magn. and Magn. Mater. **269**, 333 (2003); Mater. Sci. and Eng. **385**, 449 (2004).

Mater. Res. Soc. Symp. Proc. Vol. 1243 © 2010 Materials Research Society

Influence of Grain Interaction on the Mechanical Behavior in Shape Memory Material in Flexion Test

Jacinto Cortés[1], Fernando N. García[1], José G. González[2], Horacio Flores[3] and Alberto Reyes[4]
[1]Centro Tecnológico Aragón, FES Argón, UNAM. Edo. de México, México, C. P. 57130.
[2]Instituto de Investigaciones en Materiales, UNAM. México D.F. México.
[3]Maestría en Procesos y Materiales, Universidad Autónoma de Zacatecas, Zacatecas, México.
[4]IME, FES Aragón, UNAM. Edo. de México, México, C. P. 57130.

ABSTRACT

An experimental study of the mechanical properties of a Cu-Al-Be shape memory alloy is presented. The samples are tested in a cantilever arrangement. They consist of polycrystalline thin plates of shape memory material with a M_S near 0 °C and a monocrystalline sample with M_S near -90 °C. The measurements are made with strain gauges attached to the top side of the samples. In these conditions, strain Vs load curves are obtained. A polycrystalline sample is instrumented and tested and then it is cut into three samples for further testing. The results show a relationship between the transformation stress and the sample grain size which differs from the typical Hall-Petch relationship. The analysis of transformation plane stress diagrams shows the development of a stress component perpendicular to those induced by the applied load.

INTRODUCTION

The study of mechanical behavior of shape memory material (SMM) polycrystals is very important for developing low cost applications taking advantage of their properties efficiently. This is rather complicated because their mechanical behavior is non linear, anisotropic and depends on temperature. Additionally, microstructural factors must be considered such as grain size and shape as well as crystalline texture. Studies performed by Sakamoto and Shimizu [1] in Cu-Al-Ni polycrystalline shape memory materials subjected to simple tension tests at several temperatures have shown that the transformation stress increases with respect to the theoretical value. This expected value had been calculated on basis of the transformation temperature (MS) obtained by a differential scanning calorimetry test. These phenomena induce an effect of decreasing of transformation temperature (MS) and lead to an apparent transformation temperature (MS_{ap}).

On other hand, other investigations performed in polycrystalline Cu-Al-Ni [3, 4] and Cu-Zn-Al [2, 4] shape memory materials have shown a effects on the transformation stress subjected to simple tension tests at constant temperature. The transformation stress increases when the relative grain size decreases, this involves mainly the relationship between the grain size and the sample thickness.

The present investigation deals with polycrystalline samples and a monocrystal made of a Cu-Al-Be shape memory alloy type. Strain gages are attached to the samples for bending tests in a cantilever arrangement in order to investigate the effect of the ratio grains size/sample width on the transformation stress.

EXPERIMENTAL METHOD

The polycrystalline material is cast from a controlled environment induction furnace with a chemical composition close to Cu-Al 12%-Be 0.5% (weight percent). The ingots are hot-rolled until plates of 1 mm in thickness are obtained.

All the samples are tested by attaching strain gages (EA-06-062AQ-350, EA-06-015DJ-120 and BLH: FAE-06-12-S6) in two and four points and a bending load in a cantilever arrangement as shown in Figure 1.

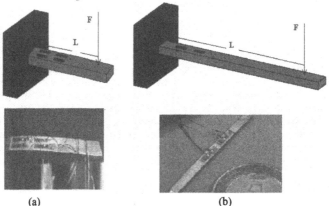

(a)   (b)

Figure 1. Load arrangement and samples instrumented with strain gages: (a) Samples M1, M2 and M5, (b) Samples M2-1 and M2-2.

The dimensions of the studied samples and the positions where the strain gages are placed are shown in Table I. After testing, sample M2 is cut longitudinally to obtain two new samples that are named as M2-1 and M2-2. This new samples are also instrumented and subjected to a bending load.

Table I. General dimensions for the samples and positions of the strain gages.

	General dimensions (mm)			Transformation temperature (° C)	Strain gages positions (mm)		Application of the load distance (L) mm	Material
	Length (L)	Width (b)	Thickness (h)		X1	X2		
M1	41.7	8	1.27	1.5	4.3	9	28	Polycrystal
M2	36	7.62	1.2	1.5	2.54	7.3	36	Polycrystal
M2-1	36	2.2	1.2	1.5	1.78	9.4	243	Polycrystal
M2-2	36	2.2	1.2	1.5	2.4	9.5	25	Polycrystal
M5	50	14	1.4	-98	3.6	14	36	Monocrystal

RESULTS AND DISCUSSION

Two examples of stress-strain curves corresponding to each strain gage placed for the samples in Table I are shown in Figure 2 where the stress is calculated using the typical solid mechanics equations for a cantilever beam. In these curves the typical superelastic loop of Shape memory materials in austenitic phase is observed [5] therefore it is assumed that a stress induced martensite transformation takes place in the samples employed in this study.

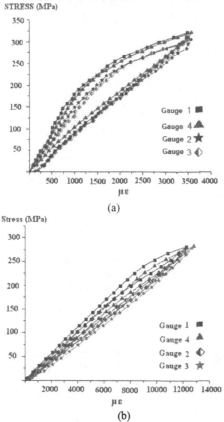

(a)

(b)

Figure 2. Stress-Strain curves and super elastic loop obtained of each strain gage placed in the samples: a) M2 b) M5

From the curves in Figure 2, the transformation stress is obtained for each curve by using the slope intersection method. With this value, the apparent transformation temperature (MS_{ap}) can be calculated and they are compared afterwards with the transformation temperature MS obtained by Differential Scanning Calorimetry in Table II. It is shown in Table II that the

117

samples M2-1 and M2-2 have a considerably small transformation stress (TS) value as compared to the M2 sample despite being the same material with identical measured MS. Table II also shows that MS_{ap} and MS have practically the same value for the monocrystalline sample (M5).

Table II. Critical stress, MS_{ap} and MS in the material for samples M2, M2-1 y M5

	MS	Strain Gauge 1		Strain Gauge 4		Strain Gauge 3		Strain Gauge 2	
		TS (MPa)	MS_{ap}	TS (MPa)	MS_{ap}	TS (MPa)	MS_{ap}	TS (MPa)	MS_{ap}
M2	1.5	183	-96	173	-98.4	173	-84.8	176	-91.3
M2-1	1.5	100	-32						
M2-2	1.5	102	-33.6	-	-	-	-	-	-
M5	-98	269.2	-111.6	260	-104.4	-	--	---	-

The transformation stress variation (this means the MS_{ap} variation shown in Table II) can be attributed to the influence of neighboring grains (or grain interactions) since this variation decreases when the sample width decreases and it disappears for the case of the monocrystal. However, it is important to clarify that the effect reported here is different to the Hall-Petch effect found by other authors for Cu-Al-Ni [3] and Cu-Zn-Al [2] in simple tension tests. In these studies the authors report a relationship between the transformation stress and the relative grain size (sample thickness/grain size rate). In our experiments the thickness is held constant (mainly in the samples M2, M2-1 and M2-2) and the load is applied by bending. Regardless of the method to apply the load (simple tension or bending), the increase of the transformation stress can be associated to grain interactions with neighbors. Thus it is reasonable to consider that these interactions can be manifested as a stress perpendicular to the normal stress induced by the applied load. In this case a plane stress state transformation diagram is particularly useful.

Figure 3 shows transformation plane stress-state diagrams for some of the main crystalline directions in the Cu-Al-Be shape memory alloy. The transformation system of this alloy reported by B. Kaouache et al. [6] is employed to build the diagrams following the procedure proposed by Bucheit et al. [7]. In Figure 3, it is clear that for crystalline orientations near $[\bar{1}11]$ and [011], the transformation diagrams show a hard distortion with respect to the transformation diagram for the crystalline orientation [001].

Employing the above criteria, Figure 4 is built and shows the transformation stress ($\sigma 1$) values corresponding to the samples: M2, M2-1 y M5 in the form of a generic transformation plane stress-state diagram for a shape memory material. In this Figure, it can be seen that the increase of transformation stress ($\sigma 1$) can be associated to the presence of a tensile stress perpendicular ($\sigma 2$) to those generated by the applied load. This is properly satisfied for sample M2-1 but the criterion is not exact for the sample M2. In this case, the results are most likely influenced by a crystalline texture in the sample which is not evaluated in the present study. Fig. 5 also shows that for sample M5, the transformation stress corresponds with the results for simple tension because in the monocrystalline sample, there no grain interactions. This is of importance because it represents an indication that it is possible to quantify the grain interactions as a stress perpendicular to those induced by the applied load.

In Figure 4, plane stress-state diagrams for the main crystalline directions are shown the corresponding Cu-Al-Be shape memory alloy. The same transformation system is used [5].

Figure 4 clearly shows that for crystalline orientations near [-111] and [011], the presence of a small perpendicular stress can induce an increase of the applied tensile stress along these directions. Nevertheless quantifying such an effect requires consideration of the crystalline texture in the samples which is beyond the scope of this investigation.

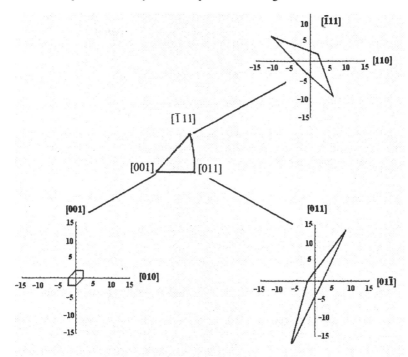

Figure 3. Transformation plane stress-state diagrams for the main crystalline directions in a Cu-Al-Be shape memory alloy.

CONCLUSIONS

A relationship between the transformation stress and the number of lateral neighbors is found in this work.

A criterion is proposed to interpret the interactions between neighboring grains as a stress perpendicular to those induced by the applied load.

ACKNOWLEDGMENTS

The authors would like to express their gratitude to Raúl Rojo Viloria, David Becerril García and Fernando Paris Delgado Gómez and the technical staff who supported the elaboration of this work: Antonio González Montaño and Alberto Higuera García.

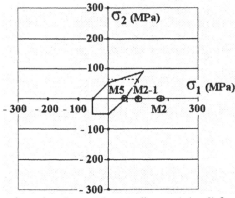

Figure 4. Generic transformation plane stress-state diagram ($\sigma1$ -$\sigma2$) for a polycrystalline shape memory material and the transformation stress values for samples M2, M2-1 and M5 plotted on the $\sigma1$ axis.

REFERENCES

1. H. Sakamoto and K. Shimizu. Experimental Investigation of Cyclic Deformation and Fatigue Behavior of Polycrystalline Cu-Al-Ni Shape memory alloys above M_S. Transactions of Japan Institute of Metals, 8, pp 592-600, (1986).

2. M. Somerday, R. J. Comstock, JR. And J. A. Wert. Effect of grain size on the observed pseudoelastic behavior of a Cu-Zn-Al shape memory alloy. Metallurgical and Materials Transactions A. 28A. pp. 2335-2341(1997).

3. G. N. Sure and L. C. Brown. The mechanical properties of refined β Cu-Al-Ni strain-memory alloys. Metallurgical Transactions A. 5ᵃ. pp 613-1621(1984).

4. S. S. Leu, Y. C. Chen, R. D. Jean. Effect of rapid solidification on mechanical properties of Cu-Al-Ni shape memory alloys. Journal of Materials Science. 27.pp 2792-2798, (1992).

5. K. Otsuka and C. M. Wayman. Introduction to shape memory materials. Shape memory materials. Cambridge University Press. pp. 1-48 (1998)

6. B. Kaouache, K. Inal, S. Berveiller, A. Eberhardt and E. Patoor. Martensitic transformation criteria in Cu–Al–Be shape memory alloy. In situ analysis. Materials Science and Engineering A. 438-440, 773 (2006)

7. T. E. Buchheit and J. A. Wert. Modeling the Effects of Stress State and Crystal Orientation on the Stress-Induced Transformation of Ni-Ti Single Crystals. Metallurgical and Materials Transactions A. 25A. pp. 2383-2389. (1994).

Mater. Res. Soc. Symp. Proc. Vol. 1243 © 2010 Materials Research Society

Inter Laboratory Comparison and Analysis on Mechanical Properties by Nanoindentation

J. M. Alvarado-Orozco[1], C. Cárdenas-Jaramillo[1], D. Torres-Torres[1], R. Herrera-Basurto[2], A. Hurtado-Macias[3], J. Muñoz-Saldaña[1,3*] and G. Trápaga-Martinez[1]

[1]Centro de Investigación y de Estudios Avanzados del IPN, Unidad Querétaro, Libramiento Norponiente 2000, Real de Juriquilla, 76230, Querétaro, México.
[2]CENAM-Dirección de Metrología de Materiales, 76241, Querétaro, México.
[3]Centro de Investigación en Materiales Avanzados, S.C., Laboratorio Nacional de Nanotecnología, Miguel de Cervantes 120, Complejo Industrial Chihuahua, 31109, Chihuahua, México.
* On sabbatical leave.

ABSTRACT

This contribution presents the results obtained by a Mexican laboratory in the *Asia-Pacific Economy Cooperation* Interlaboratory Comparison (IC) on mechanical properties by nanoindentation from 2008 using fused silica and polycarbonate as samples. Reduced modulus and indentation hardness are the parameters asked to be measured and compared. The aim for this paper is to show and to discuss the so called "indentation size effect" (ISE) on the indentation hardness of fused silica. Using the spherical formulation of the ISE model for crystalline materials, the macroscopic hardness and material length scale of fused silica are determined as (7.34 ± 0.085) GPa and (166.36 ± 14) nm, respectively.

INTRODUCTION

Nanoindentation is a well established method to characterize mechanical properties of materials. The distinguishing feature of this technique is the indirect measurement of the contact area using a correlation with the penetration depth. A typical single load nanoindentation test produces a load-penetration curve from which property values such as indentation hardness (H_{IT}) and reduced elastic modulus (E_r) can be calculated using a variety of approaches such as the Oliver and Pharr method [1]. However, irrespective of the type of analysis, issues associated with the nanometric scale (e.g. calibration, tip wear) must be accounted for. Several factors may modify nanoindentation results, being the indentation size effect, which is closely associated to the tip radius one of the most important. Due to the importance of characterizing materials in nanometric-scale, reliable measurements are considered a prerequisite for nanotechnology development. One of the most successful ways to assess and evaluate the quality of measurements is the participation in interlaboratory comparisons (IC). The *Asia-Pacific Economy Cooperation* (APEC) organized an IC in 2008 on nanoindentation. The aim of the comparison is to establish the current quality of the results obtained by different laboratories and it is conducted by the Industrial Science and Technology Working Group (ISTWG) through the NanoTechnology Research Center of Industrial Technology Research Institute (NTRC-ITRI) [2]. The IC of the APEC involved 13 laboratories from 8 different countries. The Mexican laboratory (Mex-Lab) is represented by Cinvestav–Queretaro. This paper presents the results of the IC and particularly discusses the indentation size effect (ISE) in the hardness values of fused silica. We claim, that the data scattering is associated with the degree of wear of the used indenter being however an intrinsic behavior of fused silica.

EXPERIMENTAL PROCEDURE

Interlaboratory comparison

Two samples, fused silica (FS) (8x8x3 mm) and polycarbonate (PC) (16x16x5 mm), manufactured by ASMEC Co (Advanced Surface Mechanics, Germany) are provided to the participant laboratories in order to determine and compare E_r and H_{IT}. Uniformity and homogeneity tests are performed by the *pilot laboratories* using a Hysitron instrument with a Quasi-Static method prior to deliver the samples to the participants [2]. The measurement procedure is established by NTRC-ITRI. A diamond tip with Berkovich geometry is used by all participant laboratories. The details of the procedure used by the MexLab are as follows: Indentation measurements are conducted using a Ubi1 instrument (Hysitron). The machine compliance and the area function of the tip are calibrated before the indentation measurements. A trapezoidal three-segment load function is used. The loading and unloading segments are completed over a time of 30 s irrespective of the maximum load (F_{max}). F_{max} is kept constant during 30 s. A set of 49 indentations is carried out in a symmetrically spaced matrix, where consecutive indentation imprints are separated at least 1.5 μm from each other to avoid the influence of the stress fields around the indentations. A 60 s delay is established before and after each indentation to determine the thermal drift. The mechanical properties E_r and H_{IT} are determined and submitted for the IC. The hardness is calculated according to $H_{IT}=F/A_c(h_c)$, where F is the applied load and A_c is the contact area, which is a function of the contact depth (h_c) as calculated by the Oliver and Pharr Method [1,2]. The reduced elastic modulus is calculated using equation 1.

$$\frac{1}{E_r} = \frac{1-v_i^2}{E_i} + \frac{1-v_s^2}{E_s} = \frac{2}{\sqrt{\pi}} \cdot \frac{\sqrt{A_c(h_c)}}{S},$$ (1)

There E and v are the Young's modulus and Poisson's ratio, respectively, and the subscripts i and s are associated with the indenter and sample, respectively. The contact stiffness, $S=dF/dh$ is estimated from the first part of the unloading segment of the load-penetration curve. The Z-scores method is applied to determine the consistency of the results according to equation 2.

$$Z = \frac{Result - Median}{0.7413 \cdot IQR},$$ (2)

The median is chosen as the consensus value and the Normalized Inter-Quartile Range (0.7413 · IQR)=$NIQR$ is properly used to estimate the variability between the set of results.

Depending on the $|Z| \leq 2$, $|Z| \leq 3$ and $|Z| > 3$ values, the results reported from all participant laboratories are interpreted as **satisfactory, questionable** and **unsatisfactory**, respectively.

RESULTS

Reduced elastic modulus and hardness of polycarbonate

The reported values for polycarbonate from all participant laboratories are consistent. The median and mean E_r values of PC are 3.13 GPa and (3.16 ± 0.2) GPa, respectively, with a $NIQR$ equal to 0.11. The result of the MexLab is (3.12 ± 0.02) GPa with a Z-score of -0.09

(**satisfactory**). The median and mean H_{IT} values of PC are 0.19 GPa and (0.19 ± 0.03) GPa, respectively, with a *NIQR* equal to 0.02. The reported value from the MexLab is 0.18 GPa, leading to a Z-score of 0.5, that again corresponds to a **satisfactory** result [2].

Reduced elastic modulus of fused silica

Figure 1a shows the results for the reduced elastic modulus of fused silica obtained by the fifteen participant laboratories. The reported results are in the range of 63.8 and 74.5 GPa. The median and mean values are 69.9 GPa and (70.12 ± 2.28) GPa, respectively, with a *NIQR* equal to 1.85. The literature most accepted value for the reduced elastic modulus is 69.9 GPa [3]. The obtained value from the MexLab is (69.8 ± 0.27) GPa. The corresponding Z-scores for each participant laboratory are plotted in Figure 1b. The Z-score of the MexLab is 0.05, as can be seen, the results delivered by most laboratories are **satisfactory**, except for those obtained by laboratories L09, L12 and L14 [2].

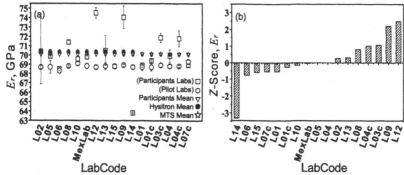

Figure 1. Summary of E_r results (a) and the corresponding Z-scores (b) of fused silica.

Indentation Hardness of fused silica

Figure 2a shows the corresponding results for H_{IT} of fused silica. The results from all laboratories oscillate between 8.2 and 10.7 GPa. The median and mean values are 9.5 GPa and (9.4 ± 0.7) GPa, respectively, with a *NIQR* equal to 0.67. The obtained value from the MexLab is (8.2 ± 0.1) GPa. Based on the obtained results the Z-scores are calculated and plotted in figure 2b taking into account all the reported data. The Z-scores obtained by all laboratories are within $2 \leq |Z| \leq 3$, clearly showing that the hardness values of fused silica obtained by the participant laboratories are highly **consistent** to each other. It is noteworthy that the corresponding Z-score of the MexLab is -1.94, which corresponds to the worst value. Irrespective of the low data scattering from the hardness measurements, our interpretation is that such deviations contain information related to the indenter geometry, particularly the tip radius of the used indenters. Such an effect has been observed in several crystalline materials [4,5,6]. For instance, important variations in the H_{IT} have been observed in functional ceramics due to the ISE and associated to the onset of plasticity as well as non-linear deformation phenomena, such as ferroelasticity [4,6,7]. One can use this concept as a starting point to explain the deviation in the hardness values of the Mexican laboratory taking into account the effect of indenter tip radius on the deformation mechanisms of fused silica. Thus, an ISE analysis of fused silica is discussed in the next section.

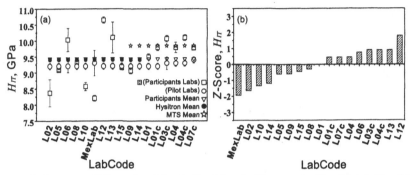

Figure 2. Summary of H_{IT} results of fused silica (a) and of Z-scores for H_{IT} of fused silica (b).

DISCUSSION

ISE is an increase of hardness with decreasing either indentation depth or indenter tip radius. This is produced by strain gradients in the indentation stress field during the indentation test [4,5,8,9,10]. The spherical formulation of the ISE model relates the contact pressure to the macroscopic hardness (H_0) and generalizes the effect, irrespective of the geometry of the indenter, by considering an equivalent tip radius according to Equation 3 [10].

$$\frac{H_{IT}}{H_0} = \sqrt{1 + \frac{R_a}{R_r}},$$ (3)

Here, H_0 is the hardness at infinite indentation depth, R_r the spherical radius of the residual surface impression and R_a is defined as a material length scale [10]. To obtain the ISE of fused silica, additional indentation tests are carried out using five Berkovich diamond tips with different grades of wear that lead to tip radius differences in the range between 250 to 860 nm. The tips are named BK1, BK2, BK3, BK4 and BK5, where the number of identification increases with the tip radius. The indenter BK4 is the one used for the IC.

To assure precision of the ISE analysis, the machine compliance and the area function are previously calibrated for each indenter following the exact procedure as for the IC. The area function and the tip radius are obtained in the range of 9000 µN-50 µN and 2000µN-50µN, respectively. The tip radius is calculated based on the projected area of the indentation imprint considering the tip radius as an equivalent sphere at a contact depth h_c in the elastic regime according to:

$$A_c = \pi a^2 = -\pi h_c^2 + 2\pi R_T h_c.$$ (4)

Where A_c, R_T and a are the projected contact area, tip and contact radius, respectively. R_T is obtained by fitting Ec. 4 to the experimental data A_c-h_c. Figure 3a shows both the experimental data and the fitting procedure in the proper range. A so called adjusted R-square statistic is used as a fitting quality factor. Table I shows the condensed results of H_{IT} and R_T for each indenter.

Table I. Summary of results that include indentation hardness (H_{IT}), tip radius (R_T) and residual tip radius (R_r) of the used indenters.

ID	H_{IT}(GPa)	R_T (nm)	*Quality factor*	R_r (nm)	*Quality factor*
BK1	8.98±0.11	258.5±13	0.821	357.8±4	0.981
BK2	8.83±0.07	281.6±14	0.975	366.2±3	0.975
BK3	8.60±0.09	321.7±9	0.956	429.0±4	0.974
BK4	8.20±0.10	447.6±3	0.988	642.7±12	0.946
BK5	7.76±0.10	885.3±9	0.998	1537.6±52	0.827

The contact radius, is calculated for each indentation test using equation 5 and used to obtain the residual tip radius (R_r) needed for the ISE analysis according to $a^2 = 2R_r h_r - h_r^2$, where h_r is the residual penetration depth after complete unloading and is given directly by the instrument. The results for R_r and their corresponding adjusted R-square values at shallow penetration depths (less than 40 nm) are shown in Table I. Finally, using the H_{IT} and R_r data and applying equation 3, the parameters H_0 and R_a are estimated [10] leading to a successfully fitting procedure using the ISE model (Fig. 3b). Thus, as shown in Fig. 3b, H_{IT} of fused silica decreases as the tip radius of the indenter increases and the results clearly indicate that ISE takes place during nanoindentation tests of fused silica. The fitting parameters H_0 and Ra are 7.34±0.085 GPa and 166.36±14nm, respectively.

The tip radii for the reported range of hardness values (8.2>H_{IT}> 10.7 GPa) reported in the IC are calculated using equation 3 leading to values in the range of 170 and 440 nm. The possibility of the participant laboratories to have been using for the IC Berkovich tips with R_T values of less than 440 nm is definitively real.

The ISE behavior of fused silica as a "source" of deviation for the MexLab hardness results has been fully demonstrated. In other words, none of the hardness measurements from the participant laboratories is wrong but are obtained using indenters of different radii. Now, due to the importance of fused silica as a reference material several questions regarding its mechanisms of deformation are now open. The onset of plasticity, where size effects play the most important role are of particular interest. There, the mechanisms of plastic deformation are still almost unknown and deserve further attention.

CONCLUSIONS

Based on the data of the APEC-IC, highly reproducible results of Young Modulus and hardness are found in fused silica and polycarbonate. Slight deviations in the results for the hardness of fused silica, varying in the range of 8.2 to 10.7 GPa are explained in terms of size effects during nanoindentation. A spherical formulation of an ISE model typically used for crystalline materials allows extrapolation of the reported results to plausible radii of the used indenters from the participant laboratories. Thus, it is proposed to include the actual radius of the indenter as an additional control parameter for future interlaboratory comparisons on nanoindentation.

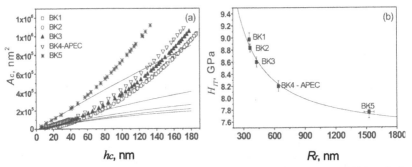

hc, nm Rr, nm

Figure 3. Determination of the tip radius of each used Berkovich indenter (a) and ISE analysis of fused silica (b).

REFERENCES

[1] W.C. Oliver and G.M. Pharr, *J. Mater. Res.*, **7**, 1564–1583, (1992).

[2] 2008 APEC ISTWG Project Interlaboratory Comparison on Mechanical Properties by Nano Indentation - *Measurement Report Draft 1*. (2008).

[3] A.C. Fischer-Cripps, *Nanoindentation*, 2–49, Berlin: Springer (2004).

[4] A. Hurtado-Macias, J. Muñoz-Saldaña, F.J. Espinoza-Beltran, T. Scholz, M.V. Swain and G.A. Schneider, *J. Phys. D: Appl. Phys.* **41**, 035407, (2008).

[5] W.D. Nix and H. Gao, *J. Mech. Phys. Solids*, **46**, 411–25, (1998).

[6] T.T. Zhu, X.D. Hou, A.J. Bushby and D.J. Dunstan, *J. Phys. D: Appl. Phys.*, **41**, 074004, 6, (2008).

[7] T. Scholz, J. Muñoz-Saldaña, G.A. Schneider and M.V. Swain, *Appl. Phys. Lett.*, **84** (16), 3055-3057, (2004).

[8] N.A Fleck, G.M. Muller, M.F. Ashby and J.W. Hutchinson, *Acta Metall. et Mater.*, **42** (2), 475-487, (1994).

[9] Q. Ma and D.R. Clarke, *J. of Materials Research*, **10** (4) 853-863, (1995).

[10] J.G. Swadener, E.P. George and G.M. Pharr, *J. Mech. Phys. Solids*, **50**, 681–694, (2002).

Mater. Res. Soc. Symp. Proc. Vol. 1243 © 2010 Materials Research Society

Synthesis and Characterization of Polyurethane Scaffolds for Biomedical Applications.

M.C. Chavarría-Gaytán[1], I. Olivas-Armendáriz.[1,2], P.E. García-Casillas,[1]A. Martínez-Villafañe[2] and C. A. Martínez-Pérez[1]
[1]Departamento de Ciencias Básicas, Instituto de Ingeniería y Tecnología, Universidad Autónoma de Ciudad Juárez, Chih. México.C.P. 32310
[2]Centro de Investigación de Materiales Avanzados S.C., Chihuahua, México, C.P. 31109

ABSTRACT

Polyurethanes are interesting materials that can be used in biomedical applications for regeneration of bone tissue. In this work the synthesis and characterization of porous polyurethanes to act as scaffold is performed by a thermally induced phase separation technique. The appropriate parameters are determined in order to obtain a porous well interconnected material. Characterization by thermogravimetric analysis (TGA) and differential scanning calorimetry (DSC) is made in order to determine the thermal stability of the material. Chemical characterization is made by Fourier transformed infrared spectroscopy with attenuated total reflectance (FTIR-ATR). The morphology of the material is observed by a field emission scanning electron microscope (FESEM) and the mechanical properties are measured by dynamic mechanical analysis (DMA).

INTRODUCTION

Engineering any human tissue requires several basic components, among them the scaffold or a delivery matrix and viable cells. Thus, the scaffold is an essential material to support initial cell growth and differentiation. Ideally, scaffolds must be made of biocompatible materials to avoid host rejection. It is also important to have a degradable matrix that provides sufficient initial strength. The matrix must be degraded over a period of time to allow the growth of regenerating tissue [1]. A great variety of technologies has been developed to produce polymeric porous scaffolds. The conventional techniques include fiber bonding, solvent casting, particulate leaching, membrane lamination, melt molding, emulsion, freeze drying and supercritical fluid technology. Nevertheless, comparison of the mechanical properties of current man made porous supports with those of bone reveals insufficient mechanical integrity. This is the reason for further research efforts in order to find materials for bone tissue [2, 3].
Common support materials used in tissue engineering are produced from natural or synthetic materials. Polymers, polysaccharides, polyesters, hydrogels or thermoplastic elastomers, and ceramic assets like calcium phosphates [4, 5]. An interesting material is the polyurethane (PU) that has many useful attributes in tissue engineering such as durability, elasticity, fatigue strength and tolerance in the body during the treatment. For that reason, polyurethanes are considered excellent candidates for medical devices and biomedical applications, although most applications have been limited to non-degradable matrices [6]. In this study, thermally induced phase separation (TIPS) is used to obtain a scaffold with good mechanical properties, an interconnected porous structure, and to control porosity, morphology, bioactivity and degradation rates that could be used for biomedical applications.

EXPERIMENT

All materials have been purchased from Sigma Aldrich and used as received. The PU samples are prepared by a methodology previously reported with a slight modification [7]. Polycaprolactone diol and polycaprolactone triol are dissolved in 1,4 dioxane with c-hexane as a co-solvent (Table I). Once the reagents are dissolved; 1,6 hexanedisocyanate is added having 0.5 weight % dibutylin dilaurate as catalyst. Subsequently, the homogeneous solutions are quenched to -15 °C and then the solvent is extracted for 48 h in a freeze drying system (Labconco FreeZone 2.5). Finally, the sample is cured at 60°C for 24 h.

Table I. Specifications of the compositions of PUs.

Sample	Polyol a (g)	Triol b (g)	Dioxane (%vol)	Hexane (%vol)	Water (%vol)
PU1	PCLd-1 5	PCLt-1 0.8	87	13	0
PU2	PCLd-2 5	PCLt-1 0.8	87	13	0
PU3	PCLd-1 5	PCLt-1 0.8	87	0	13
PU4	PCLd-1 5	PCLt-1 0.5	87	13	0
PU5	PCLd-1 5	PCLt-1 0.5	93	7	0
PU6	PCL-1 5	PCLt-1 0.5	87	13	0

a. Diol molecular weights: PCLd-1, 1250; PCLd-2, 2000
b. Triol molecular weight: PCLt-1, 900.

Morphological characterization is made in a field emission scanning electron microscope (FESEM) Jeol JSM7000F. The thermal characterization of the materials are made by means of calorimetric measurements in a simultaneous TGA/DSC system (SDT-Q600 TA instruments). The samples are cut in small pieces and heated at a constant rate of 10°C/min from room temperature up to 750 °C under a flow of nitrogen (10 mL/min).

The infrared spectra are acquired in a Perkin Elmer spectrometer, model spectrum GX) using attenuated total reflectance (ATR) in a SMITHS system (Durasample IRII model) that has a diamond window of 2 mm of diameter.

The mechanical properties are measured in a TA Rheometrics Analyzes System RSA III in the frequency range of 1×10^{-6} to 80 Hz. Stress-strain tests are made at 40°C with a deformation rate of 0.001m/s.

DISCUSSION

128

The thermal stability of PUs is evaluated from the TGA data derived from the thermogravimetric analysis under N_2 atmosphere (Figure 1 and Table II). The degradation of PU takes place in two steps. The first step of degradation can be attributed to an inverse reaction of polyaddition. It leads to formation of isocyanate and alcohol groups and takes place in the temperature range around 160-400 °C. This first step presents a maximum weight loss rate that is determined from the peak of the first derived of each phase of degradation (Td_{max}) as can see in Table II. In order to analyze these results it is important to establish comparative parameters. First, the thermal stability can be improved increasing the molecular weight of polyol according to the results of samples PU1 and PU2 where the Td_{max} is increased by 31 ° C, changing from 332° C to 363° C. Secondly, if the proportion of triol in the sample is increased, then the Td_{max} increased 17° C as it can be seen in samples PU1 and PU6. On the other hand, according with the results in PU3 and PU4, keeping the concentration of dioxane and water constant and increasing the quantity of triol increase Td_{max}. Furthermore, if it is compared the effects of using hexane or water, in samples PU1 and PU3, then it is possible to see that the use of water increase the Td_{max} for 13° C. Finally, the second step of degradation in the PU is carried out between 400 to 600 °C that it is related to the oxidation of residual material. These results are consistent with previous investigations [8], In summary, the results suggest that increasing the molecular weight and the proportion of triol have more influence in the thermal properties of PU than quantity of the solvents.

All of the IR spectra present the characteristic peaks of PU. In Figure 2, the IR spectrum of the PU4 is presented as an example. The absorption band at 3400 cm^{-1} corresponds to the stretching mode of -NH groups, and the absorption band between 2800-3000 cm^{-1} is associated with asymmetric and symmetrical -CH$_2$ groups. The band at 1800 cm^{-1} corresponds to the presence of the -C=O groups. The absorption bands between 1500-1600 cm^{-1} corresponds to stretching of -NH, +-C-N+C-C, which are sensitive to formation of the chain and the intermolecular hydrogen bonding. Finally, the absorption band at 1100 cm^{-1} is due to stretching of C-O-C groups. These results confirm the typical chemical structure of Polyurethane and they are consistent with previous investigations [9, 10].

Table II. Degradation temperatures and residual weight % of the first step of PUs.

Sample	Td_{on} (°C)	Td_{max} (°C)	%weight remaining at Td_{max}	Td_{final}	%weight remaining at 400°C
PU1	213.0	332.7	61.21	401.7	21.1
PU2	179.6	363.8	44.21	466.2	12.8
PU3	160.0	345.3	38.1	404.0	11.5
PU4	215.3	363.1	42.3	460.4	11.7
PU5	216.5	351.1	44.9	417.8	8.2
PU6	160.0	349.9	48.5	393.7	24.3

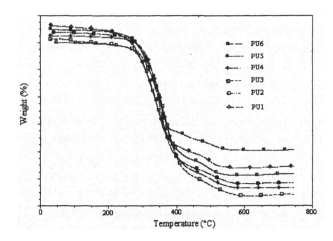

Figure 1. TGA of the PU samples under nitrogen atmosphere .

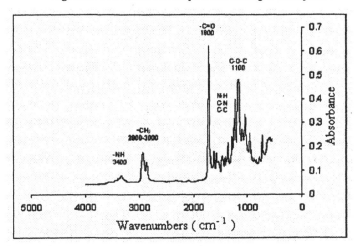

Figure 2. FTIR spectrum PU4 (PCLd-1, PCLt-1) and solvent relation (83:13:0)

The images of the FESEM of the PU4 and PU5 are shown in the Figure 3. An open microporous structure with interconnected pores can be observed. It is well known that during a freeze–drying process, thermally induced phase separation produces an interconnected porous microstructure in the polyurethane materials [3]. Pore size is in the range of 100 to 120 μm, which is appropriate for the vascularization process required for bone tissue regeneration.

130

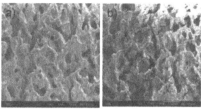

Figure 3. SEM micrographs of: a) PU4 , and b) PU5 .

The compressive mechanical properties of PU2, PU3 and PU5 are shown in Figure 4. They are important for practical applications because there is a close relationship between dimension-maintaining ability in the scaffolds. These results exhibit a significant increase in the strength range of the PU5 compare with the PU3 and PU2. The PU5, which is prepared with the greatest amount of PCL triol using a 93:7:0 mixture of dioxane, hexane and water has the highest strength range and the lower strain range (highest Young´s modulus, 2.3 KPa). On the other hand, the lowest strength (1.044x10⁵ Pa) and largest elongation (49,64%) is obtained in material PU3 that is prepared using water as co-solvent and the PCL diol and triol with the smallest molecular weights. These results suggest that future research could analyze the effect of solvents, varying the molecular weight, keeping constant the quantities of solvents to optimize the mechanical properties of the composite.

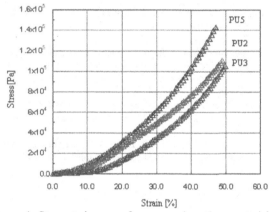

Figure 4. Stress-strain curves for some polyurethane materials

131

CONCLUSIONS

It is possible to obtain PU scaffolds with appropriate morphology that can be used for bone tissue engineering. According to the mechanical properties obtained, these materials can be used for spongy bone regeneration.

ACKNOWLEDGMENTS

The authors are very grateful to CONACYT México for their financial support and to Mónica E. Mendoza Duarte by her support during the mechanical properties characterization.

REFERENCES

[1] S. Agarwal, R. Gassner, N.P. Piesco, S. R. Anta in *Tissue Engineering and Biodegradable Equivalentes*, edited by K. Lewandroski, D.L. Wise, D.J.Trantolo, J.D. Gresser, M.J. Yaszemski and D.E. Altobelli, Marcel Kekker, Inc. (2002), pp.123-135.

[2] K. Rezwan, Q.Z. Chen, J.J. Blaker, A.R. Boccaccini: Biomaterials 27 p. 343 (2006)

[3] A. Martínez, P.E. Casillas, A. Martínez Villafañe and J. Romero-García, Journal of Advanced Materials Sp. Ed. 1, 5 (2006).

[4] M.R: Williamson and A.G.A. Coombes, Biomaterials, 25, 459 (2004).

[5] D.W. Hutchmacher, Biomaterials, . 21, 2529 (2000)

[6] H. Choand J- An., Biomaterials, 27, 544 (2006).

[7] J.H. deGroot J.H., A.J:Nijennhuis.,P. Bruin., A.J.Pennings., R.P.H. Veth., J. Klompmaker. and H.W.B. Jansen, Colloid and Polymer Science, 268, 1073 (1990)..

[7] D.K. Chattopadhyay, A.K. Mishra and B. Sreedhar, K.V:S.N: Raju: Polymer Dedradationand Stability, 91, 1837 (2006).

[8] K. Gorna, and S. Gogolewsli, Polymer Degradation and Stability, 79, 113 (2002). .

[9] M. Berta, C. Lindsay, G. Pansand G. Camino: Polymer Degradation and Stability 91, 1179 (2006).

[10] S. Duquesne, M. Le Bras, S.Bourbigot, R. Delobel, G. Caminio, B. Eling, C. Lindsay and T. Roels: Polymer Degradation and Stability,. 74 , 493 (2001)..

Mater. Res. Soc. Symp. Proc. Vol. 1243 © 2010 Materials Research Society

Glass-Ceramics of the Wollastonite - Tricalcium Phosphate-Silica System

Jorge López-Cuevas, Martín I. Pech-Canul, Juan C. Rendón-Angeles, José L. Rodríguez-Galicia and Carlos A. Gutiérrez-Chavarría
CINVESTAV-IPN Unidad Saltillo, Ramos Arizpe, 25900 Coah., México

ABSTRACT

Glass-ceramics based on hypo-eutectic (GC1) and hyper-eutectic (GC2) compositions of the Wollastonite (W, $CaSiO_3$) - Tricalcium Phosphate [TCP, $Ca_3(PO_4)_2$] binary system, which are saturated with SiO_2 during the glass melting stage, are synthesized by the petrurgic method, using cooling rates of 0.5, 1 or 2°C/h. All synthesized materials are subjected to *in vitro* bioactivity tests using Kokubo's Simulated Body Fluid (SBF). Primary α-Cristobalite is formed in all cases. Metastable Apatite [Ap, $Ca_{10}(PO_4)_6O$] and W phases are additionally formed, in general, in the GC1 glass-ceramics, as well as in the GC2 material obtained at a cooling rate of 0.5°C/h. However, at faster cooling rates, TCP is formed instead of Ap phase in the latter composition. During the bioactivity tests, a hydroxyapatite [HAp, $Ca_{10}(PO_4)_6(OH)_2$]-like surface layer is formed in all materials. It is proposed that GC2 glass-ceramics cooled at a rate of 1°C/h have the potential to show good *in vivo* osseointegration properties.

INTRODUCTION

Over the last ten years, the eutectic composition (60wt% W-40wt% TCP) of the Wollastonite (W, $CaSiO_3$) - Tricalcium Phosphate [TCP, $Ca_3(PO_4)_2$] system, has attracted a great deal of attention in the field of biomaterials, mainly due to the development of the Bioeutectic® material [1,2]. This is the first designed bioceramic with the ability to develop an *in situ* interconnected porous hydroxyapatite [HAp, $Ca_{10}(PO_4)_6(OH)_2$]-like structure that mimics porous bone, in contact with physiological fluids, which confers to it excellent osseointegration properties. Its irregular lamellar eutectic microstructure, constituted by alternating radial lamellae of W and TCP phases, is obtained by slow solidification through the eutectic temperature (1402 ± 3°C) of the binary system. Thus, nucleation and growth of the W and TCP crystals occur during slow cooling from the molten state. This procedure is called "petrurgic method" [3,4]. When Bioeutectic® interacts with a physiological fluid, W is dissolved and TCP is pseudomorphically transformed into HAp-like phase, forming an interconnected porous structure. Lastly, a dense HAp-like layer is formed by precipitation on the outer surface of the material. The outstanding characteristics of Bioeutectic® seem to have inhibited the interest of the researchers to study non-eutectic compositions of the W-TCP system, which could be also useful as biomaterials. In the present work we synthesized, by using the petrurgic method, glass-ceramics based on hypo-eutectic and hyper-eutectic compositions of the W-TCP binary system, which are saturated with SiO_2 during the glass melting stage. The *in vitro* bioactivity of all synthesized materials is evaluated.

EXPERIMENTAL DETAILS

Reagent-grade $CaCO_3$, SiO_2 and $(NH_4)_2PO_4$ raw materials are used. One hypo-eutectic and one hyper-eutectic compositions of the W-TCP binary system are considered. The hypo-eutectic binary composition is 70wt% W-30wt% TCP, containing (wt%) 36.2% SiO_2, 50.07% CaO and

13.73% P_2O_5. The hyper-eutectic binary composition is 50wt% W-50wt% TCP, containing (wt%) 25.89% SiO_2, 51.26% CaO and 22.88% P_2O_5. For the synthesis of the glass-ceramics, raw material mixtures of stoichiometric composition are prepared. Then, the mixtures are melted for 2h in fused quartz crucibles, using a high-temperature electric furnace (Lindberg/blue M), at a temperature 100°C above the liquidus temperature of each material. This is ~1440°C and ~1465°C for the hypo-eutectic and the hyper-eutectic binary compositions, respectively. Subsequently, the molten materials are cooled down at a rate of 3°C/min until a temperature 10°C above the eutectic point (1402 ± 3°C) is reached. This is followed by further cooling at a rate of 0.5, 1 or 2°C/h, until a temperature 10°C below the eutectic point is reached. Then, the furnace is switched off and the samples are allowed to cool down to room temperature inside it, in a natural way. This procedure is based on that employed for the synthesis of Bioeutectic® [1,2].

During the glass melting stage, a large amount of SiO_2 is dissolved into the molten material from the fused quartz crucibles. Assuming that the SiO_2 saturation concentration is reached during this stage, it is estimated, based on the W-TCP-SiO_2 compatibility triangle within the SiO_2-CaO-P_2O_5 ternary phase diagram [5,6], that the melt with initial hypo-eutectic binary composition achieves a final SiO_2 concentration of ~50wt%, while the melt with initial hyper-eutectic binary composition reaches a final SiO_2 concentration of ~37wt%, at the glass melting temperatures of 1540°C and 1565°C, respectively. Both SiO_2 saturation concentrations fall within the primary crystallization field of this phase in the SiO_2-CaO-P_2O_5 ternary diagram. The glass-ceramics which are derived from the SiO_2-saturated hypo-eutectic and hyper-eutectic binary compositions are hereinafter denominated as GC1 and GC2, respectively.

All materials are soaked in Kokubo's Simulated Body Fluid (SBF) [7] for 7, 14 or 21 days, at pH=7.4 and 37°C, in an incubator muffle (Fisher Scientific 637D). All materials are also characterized by Scanning Electron Microscopy (Philips XL30 ESEM) and X-Ray Diffraction (Philips XPERT XRD, CuKα radiation), before and after the bioactivity tests.

DISCUSSION

Evolution of phases

In the case of the CG1 glass-ceramics, the XRD study reveal the formation of a large amount of metastable Apatite [Ap, $Ca_{10}(PO_4)_6O$]. It remains relatively invariant, independently of cooling rate, with a small amount of W formed, except for the cooling rate of 2°C/h. TCP is not detected in any of these materials (Fig. 1 (I)). Similar results are obtained for the CG2 glass-ceramics for a cooling rate of 0.5°C/h, with a slightly higher proportion of W in this case. However, when the latter composition is cooled at a rate of 1 or 2°C/h, a large proportion of TCP is formed, instead of Ap, with a slight increment in the relative proportion of W (Fig. 1 (II)). This is determined in a qualitative way, based on the relative intensities of the phases in the XRD patterns. Thus, formation of metastable Ap takes place instead of TCP in those materials with the lowest content of CaO and P_2O_5 and with the highest SiO_2 contents, at all employed cooling rates. On the contrary, for the highest CaO and P_2O_5 and lowest SiO_2 contents, an increase in cooling rate from 0.5 to 1 or 2°C/h favors the formation of TCP, instead of Ap. Recent studies [8] have shown the formation of domains in the structure of glasses of the SiO_2-CaO-P_2O_5 system, having a chemical composition similar to that of Ap. They act as nucleation centers for the crystallization of this phase. Lastly, additionally to W, Ap and TCP, in all materials a considerable amount of SiO_2 is also detected as α-Cristobalite.

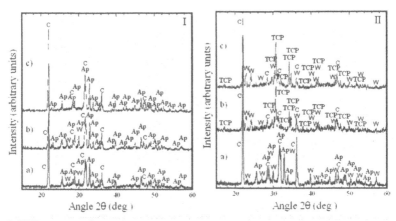

Figure 1. XRD patterns of: (I) CG1 and (II) CG2 glass-ceramics crystallized at cooling rates of: (a) 0.5, (b) 1 and (c) 2°C/h. Ap = Apatite, C = α-Cristobalite, TCP= Tricalcium Phosphate and W= Wollastonite.

Microstructural analysis

α-Cristobalite is observed on the SEM, for all materials and cooling rates employed, as ellipsoidal or globular particles with a size of ~25-300µm, and they are homogeneously distributed in the glass-ceramic matrix. Since it is known [9] that SiO_2 possesses good intrinsic bioactivity and biocompatibility properties, thus, it is assumed that the presence of these particles cannot affect negatively such properties. The globular morphology of α-Cristobalite indicates that this is the first phase crystallizing on cooling from the molten state (primary phase) [5]. Thus, it is justified to assume that a final SiO_2 concentration close to the corresponding saturation limit is achieved during the melting stage for both studied compositions. In the case of the GC1 glass-ceramics, primary α-Cristobalite, secondary W and W-TCP-SiO_2 ternary eutectic can be expected according to the SiO_2-CaO-P_2O_5 ternary phase diagram [5,6]. In the case of the GC2 glass-ceramics, almost the same phases can be expected, but now TCP would be the secondary phase.

Figure 2 shows SEM images of samples of both compositions cooled at different rates (0.5, 1 and 2°C/h). In the case of the GC1 glass-ceramics, the microstructures are similar at all cooling rates employed. The phases detected are: (A) ternary eutectic formed by Ap, W and α-Cristobalite phases, which becomes finer with increasing cooling rate, and (B) matrix corresponding to W phase (not present in the samples cooled at 2°C/h). All phase identifications are based on the XRD and SEM/EDS results (see Fig. 1 and Table I). In the case of the GC2 materials, the same phases observed for the GC1 materials are obtained for a cooling rate of 0.5°C/h. However, when the cooling rate is increased to 1 or 2°C/h, instead of the ternary eutectic we observe a new phase whose morphology is either globular (C) or elongated (D). This phase corresponds to TCP. Ap is not present in these samples. Thus, probably the high content of SiO_2 in the samples that are cooled at 0.5°C/h promotes the formation of metastable Ap, instead of TCP, which is formed when faster cooling rates are used.

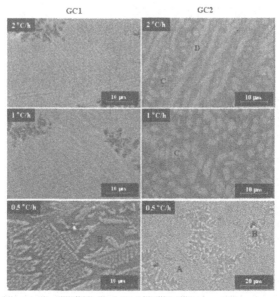

Figure 2. SEM backscattered electron images of synthesized GC1 and GC2 materials. A = Ternary eutectic, B = W phase, C and D = TCP phase. The cooling rates are indicated.

Table I. Results of SEM/EDS and BEI analysis for the observed phases.

Assumed phases		SEM/EDS chemical analysis (wt. %)	wt. av. Z
α-Cristobalite (SiO_2)		99.7% SiO_2 (D zone in Fig. 3)	10.80
W ($CaSiO_3$)		52.5% SiO_2, 2.9% P_2O_5, 44.6% CaO (B zone in Fig. 2)	13.59
TCP [$Ca_3(PO_4)_2$]		13.4% SiO_2, 40.1% P_2O_5, 46.5% CaO (C and D particles in Fig. 2)	14.05
Ternary eutectic	W ($CaSiO_3$)	30.2% SiO_2, 17.2% P_2O_5, 52.6% CaO (A zone in Fig. 2)	13.59
	TCP [$Ca_3(PO_4)_2$]		14.05
	Ap [$Ca_{10}(PO_4)_6(OH)_2$]		14.06

It is well-known that Backscattered Electron Imaging (BEI) on the SEM produces contrast by taking advantage of differences in the weight average atomic number (wt. av. Z) of the phases present in the sample [10]. The brightness of a phase increases with increasing wt. av. Z due to the generation of a larger amount of backscattered electrons from that phase. This phenomenon can allow us to distinguish one phase from another one, based on the compositional differences existing among them. Lee and Rainforth [10] give a procedure for the calculation of the wt. av. Z of any phase. This method is used to calculate the wt. av. Z values shown in Table I for the phases observed in the present work. As can be seen, the brightness of our phases should increase in the order α-Cristobalite $\rightarrow$ W $\rightarrow$ TCP, Ap. This agrees with our phase identification (see Figs. 2 and 3). Lastly, it is worth mentioning that since TCP and Ap phases have very similar values for their wt. av. Z, thus, their brightness are also very similar, and they cannot be distinguished from one another in this way within the ternary eutectic.

In Vitro bioactivity tests

A HAp-like layer, with a thickness of ~5-20µm, is formed at the surface of all samples subject to the *in vitro* bioactivity tests (Fig. 3). For both compositions studied, the Ca/P molar ratio increases in this layer with decreasing cooling rate as well as with increasing soaking time in the SBF, until a value of ~1.60 is reached. See Table II.

Table II. Results of EDS analysis of HAp-like layer formed at the surface of GC1 material, after soaking in SBF for several periods of time (the estimated accuracy of the data is ±5%).

Soaking time (days)	Cooling rate (°C/h)	Wt. %					Ca/P molar ratio
		Ca	P	Mg	Si	O	
7	0.5	47.85	24.07	---	0.89	27.19	1.54
14	0.5	48.28	23.34	0.56	0.67	26.39	1.6
21	1	44.08	23.30	---	---	32.62	1.46
21	2	39.57	21.57	1.09	---	33.71	1.42

In the GC2 glass-ceramics cooled at 1 or 2°C/h, two different zones are developed at the sample surface during the bioactivity tests: (A) dense HAp-like layer, and (B) globular particles with a size of ~2µm. Zone (C) is the unaffected region of the glass-ceramics. For both cooling rates, (D) corresponds to α-Cristobalite. For the cooling rate of 1°C/h, an additional (B') region, similar to zone (B), is detected. Based on these results, it is deduced that GC2 glass-ceramics cooled at 1°C/h and Bioeutectic® interact with SBF in a similar fashion [11]. In the case of our material, zone (B) is originated during the early stages of interaction with the SBF; its SiO_2-rich dark matrix is formed by partial lixiviation of Ca contained in W [12]. Later on, after a sufficiently long soaking time in SBF, complete dissolution of W is achieved, living a porous interconnected structure in that region, which is then filled up with HAp-like material precipitated from the SBF. This is assumed to give rise to the formation of zone (B'). Simultaneously to the dissolution of W, the globular particles (B), which correspond initially to TCP, undergo a pseudomorphic transformation into a HAp-like phase with slightly different composition with respect to zone (A). Lastly, a dense HAp-like layer, zone (A), is formed by precipitation from the SBF. These results suggest that the GC2 material cooled at 1°C/h has potentially good *in vivo* osseointegration properties.

CONCLUSIONS

In the case of the GC1 glass-ceramics, metastable Ap plus a small amount of W are generally formed. In the case of GC2 glass-ceramics, the same phases are obtained for a cooling rate of 0.5°C/h; however, TCP is formed, instead of Ap, at cooling rates of 1 or 2°C/h. Primary α-Cristobalite is formed in all cases. During the *in vitro* bioactivity tests, the formation of a HAp-like surface layer, with a thickness of ~5-20µm, is obtained in all cases. The Ca/P molar ratio increased in this layer with decreasing cooling rate as well as with increasing soaking time in the SBF. The GC2 glass-ceramics obtained using a cooling rate of 1°C/h showed a behavior similar to that of Bioeutectic® in contact with SBF. Thus, this material has potentially good *in vivo* osseointegration properties.

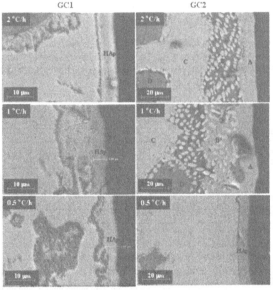

Figure 3. Cross-sectional SEM micrographs (backscattered electron images) of synthesized GC1 and GC2 materials, after 21 days of soaking in SBF.

REFERENCES

1. P.N. de Aza, F. Guitian, and S. de Aza, *Acta Metall.* **46**, 2541 (1998).
2. P.N. de Aza, F. Guitian, and S. de Aza, *Biomaterials* **18**, 1285 (1997).
3. J.M. Rincón, *Polym-Plast. Technol. Eng.* **31**, 309 (1992).
4. I. De Vicente-Moreno, P. Callejas, and J.M. Rincón, *Bol. Soc. Esp. Ceram.* V. **3**, 157 (1993).
5. R.L. Barrett, and W.J. McCaughey, *Am. Mineral.* **27**, 680 (1942).
6. J. Wojciechowska, J. Berak, and W. Trzebiatowski, *Rocz. Chem.* **30**, 751 (1956).
7. T. Kokubo, *Acta Mater.* **46**, 2519 (1998).
8. J. Pérez-Pariente, F. Balas, and M. Vallet-Regi, *Chem. Mater.* **12**, 750 (2000).
9. C.Q. Ning, J. Mehta, and A. El-Ghannam, *J. Mater. Sci. Mater. Med.* **16**, 355 (2005).
10. W.E. Lee and W.M. Rainforth, "Ceramic Microstructures: Property Control by Processing", (Chapman and Hall, 1994), pp. 133.
11. P.N. de Aza, Z.B. Luklinska, M.R. Anseau, F. Guitian, and S. de Aza, *J. Microsc.* **189**, 145 (1998).
12. P.N. De Aza, Z.B. Luklinska, M.R. Anseau, F. Guitian, and S. De Aza, *J. Dent.* **27**, 107 (1999).

Mater. Res. Soc. Symp. Proc. Vol. 1243 © 2010 Materials Research Society

Synthesis and Mechanical Characterization of Aluminum Based Composites Prepared by Powder Metallurgy

I. Estrada-Guel[1,2], J. L. Cardoso[2], C. Careño-Gallardo[1,2], J. M. Herrera-Ramírez[1] and R. Martínez-Sánchez[1].

[1] Centro de Investigación en Materiales Avanzados (CIMAV), Laboratorio Nacional de Nanotecnología, Miguel de Cervantes No. 120, CP 31109, Chihuahua, Chih., México.
[2] Universidad Autónoma Metropolitana, Depto. de Materiales, Av. San Pablo # 180, Col Reynosa-Tamaulipas CP 02200, México, D. F.

ABSTRACT

Aluminum-based composites prepared from pure Al powder and previously Cu metallized graphite are fabricated by a solid state route and are characterized by X-ray diffraction and scanning electron microscopy in order to follow their microstructural evolution. Composites are processed using powder metallurgy technique in order to obtain cylindrical samples to carry out mechanical testing. Microstructural and mechanical characterizations reveal that, by milling, a homogeneous dispersion of insoluble particles into the Al matrix is obtained; this produces an important improvement in hardness and strength with respect to an un-milled sample. Milling intensity and particle concentration have an important effect on the mechanical properties of the synthesized composites.

INTRODUCTION

Aluminum and its alloys have a wide diversity of industrial applications because of their light weight and corrosion resistance. However, their low stiffness, yield strength and resistance to wear and tear sometimes limit their use. Al-based composites are excellent alternatives to overcome these disadvantages, since they are cheap compared with other low density alloys (such as Mg or Ti), and have excellent performance [1]. On the other hand, graphite (C_g) has been recognized as a high strength, low density material. Because of its high strength to mass ratio [2] along with its excellent structural stability and mechanical performance at high temperatures, graphite has been used as a reinforcement material in polymer based composites [3]. Although the preparation of such composites by melting and casting routes is the most economical, it is associated with problems related to the poor wetting of ceramic particles [2, 4]. Therefore, the introduction and retention of graphite particles in molten Al is difficult, resulting in inhomogeneous distribution, inadequate Al/C_g bonding and formation of porosity at the matrix/C_g interface [5]. Powder metallurgy (PM) is a technology capable of providing competitive components at low cost with high material efficiency [6]; basically it consists of mixing elements or alloy powders, compacting the mixture in a die, and sintering the compacts at just below their melting point in a controlled-atmosphere furnace to bond the particles. However, raw materials only consist of simple powder mixtures. Powder mixing is a critical step [7] since it controls the distribution of particles and porosity, which influence the composite mechanical behavior. Some variables like reinforcement size-shape and type of matrix can induce agglomeration; this can be a cause of low performance [8]. If the reinforcement particles are homogeneously distributed [1] using mechanical milling (MM) in the first stage, then a good microstructural component distribution can be achieved with a decrease in their grain and particle size [9] together with an increase of the mechanical response. Additionally, solid-state processing minimizes reactions between matrix and reinforcement, which can enhance the bonding between reinforcement particles and matrix [10].

EXPERIMENTAL PROCEDURE

The initial powders (raw materials) have the following characteristics, Al from Alfa Aesar (99.5% purity and -325 mesh), Cu (99% purity and -325 mesh) and graphite (99.9% purity and -20+84 mesh). The reinforcement material i.e., metallized graphite (MG) is prepared by milling a mix of graphite-copper with 15 at.% of Cu, using a SPEX 8000M device in Ar atmosphere during 4h. The Al based composite synthesis is performed by mixing Al powder with MG particles in concentration of 0.0, 0.5 and 1.0 wt.% as shown in Table I. Then, the as-mixed powders are milled in a ZOZ CM01 Simoloyer mill for 4 milling intervals (1, 2, 4 and 8h) under Ar atmosphere. Milling device and media are made of stainless steel. Methanol is used as a process control agent to avoid excessive aluminum agglomeration. Pure Al samples (without Cu-MG addition), milled and un-milled are used as reference material for comparison proposes.

Table I. Composites nomenclature.

Sample	Milling Intensity [h]				
	0	1	2	4	8
Pure Al	P	P1	P2	P4	P8
Al + 0.5 wt.% Cu-MG	50	51	52	54	58
Al + 1.0 wt.% Cu-MG	100	101	102	104	108

In order to measure the crystallite size and lattice strain of particles, milled samples are characterized with a Siemens D5000 diffractometer. Also, a portion of composite samples are prepared using standard metallographic techniques to accomplish microstructural observations with a SEM model JEOL-JSM 7401F.

Samples are uniaxially cold pressed at 950 MPa and sintered for 3h at 823 K. Density measurements are performed according to Archimedes' method. Hardness tests are executed in sintered samples with a Wilson Rockwell hardness tester (average of five measurements is considered) in Rockwell F scale and converted to Brinell. Compression tests are carried out in an Instron universal tester at constant displacement rate of 0.0333 mm/sec, yield stress is measured at the elastic limit and maximum stress is measured at an arbitrary condition of 20% strain.

DISCUSSION

X-Ray Diffraction (XRD) characterization. Figure 1a shows some XRD patterns as a function of milling time together with magnified views of the principal Al (111) diffraction peak. Table 1 shows the nomenclature in use. It is evident that the diffraction peaks are broadening as the milling time is increased. This is induced by the stress and grain size refinement of the powder particles [11]. After 2h of processing, the peaks of all samples present a small shift to lower 2θ values, denoting an increment in the lattice parameter. However, with further milling time, the peaks return to their original position. These phenomena can be explained by the increase in the vacancy density and their subsequent relaxation, which has been reported elsewhere [12].

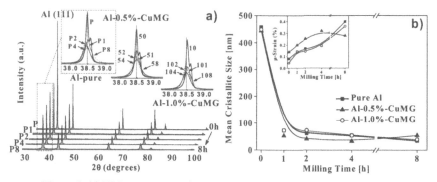

Figure 1. (a) XRD patterns and (b) Crystallite size and μ–strain of composites.

Figure 1b shows that Al-0.5%Cu-MG samples present an important variation in grain size and strain, compared with un-doped and milled samples; meanwhile Al-1.0%Cu-MG composites have a similar behavior compared with pure Al. These effects are a consequence of severe plastic deformation [13].

Morphological analysis. Since powders size distribution decreases with milling, Cu-MG particles are fractured and embedded onto the matrix surface (Figure 2). With intense milling, the surface fractures again and a new surface is exposed and covered by free particles. This process is repeated many times during processing and then the fragmented particles are captured by welding particles and confined to welding lines, gradually developing a lamellar structure. A repeated fracture-convolution process results in a uniform distribution of the Cu-MG and good bonding between layers that, contribute to increase the compression strength [14].

Figure 2. SEM images and EDS analysis on Cu-MG particle (bright spot) embedded between Al matrix layers. The particles are dispersed in the matrix and present a high concentration of C and Cu (EDS analysis). Even though the high solubility of Cu in Al, particles remain insoluble and these present nanometric size and are grouped in a cluster form.

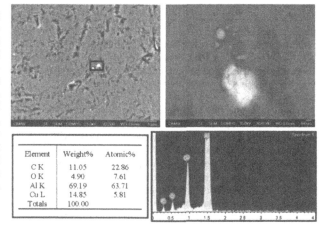

Element	Weight%	Atomic%
C K	11.05	22.86
O K	4.90	7.61
Al K	69.19	63.71
Cu L	14.85	5.81
Totals	100.00	

141

Table II. Results of compression tests $[Kg_F/mm^2]$ and hardness measurements in the sintered composites.

Sample	σ_y	σ_{max}	Brinell H.
P	7.12 ± 0.05	17.71 ± 0.04	-----
P1	10.17 ± 0.07	24.31 ± 0.22	-----
P2	14.91 ± 0.16	31.05 ± 0.33	Under 55 HB
P4	13.46 ± 0.18	27.18 ± 0.21	-----
P8	9.60 ± 0.44	30.71 ± 0.15	-----
50	6.84 ± 0.05	18.24 ± 0.04	-----
51	13.01 ± 0.31	29.72 ± 0.20	Under 55 HB
52	17.50 ± 0.18	35.71 ± 0.14	55 ± 1
54	20.62 ± 0.01	41.34 ± 0.24	66 ± 1
58	20.66 ± 0.05	33.87 ± 0.13	63 ± 1
100	6.58 ± 0.28	18.31 ± 0.20	-----
101	14.72 ± 0.25	27.02 ± 0.63	-----
102	17.14 ± 0.23	36.17 ± 0.14	59 ± 1
104	14.90 ± 0.17	32.95 ± 0.22	Under 55 HB
108	10.29 ± 0.92	35.20 ± 0.13	Under 55 HB

Densification. Figures 3a-b show some density determinations in composites. For comparison purposes between samples, the measured density is divided by the theoretical density for each sample composition in order to get the densification percent shown in these figures. It is evident that, low milling times (1h) produce samples with higher densities (Figure 3). This is characteristic of spherical morphologies where powders have a good mobility and they pack randomly and densely [15,13]. On the contrary, un-doped samples after 4h of milling show a marked decrease in their densification. Milled flattened particles yield a low densification due to high friction and bridge formation. In this way, laminar morphology of the longer-time milled powders induces poorer packing and consequently low density values [7]. Another cause of this effect can be attributed to the hardening effect of the milling, which increases the powders hardness and consequently limits their plastic deformation response [16]. These phenomena are important in samples with longer milling times (2 to 8h), where milling has an adverse effect on density. Meanwhile, Al-1.0%Cu-MG samples maintain their density and Al-0.5%Cu-MG have an increment of densification. Densification depends on milling intensity and reinforcement concentration, reaching an optimal with low additive concentration and 1-2 h of milling.

Mechanical properties of sintered composites. Figure 4a shows the average of three stress-strain curves from pure Al samples milled at five time intervals. It is clear, that milled samples present a better mechanical performance as compared with un-milled Al, due to work hardening [17] and grain size reduction [13]. Table II shows measurements of mechanical properties for all samples, it is evident that milling induces an important increase on the yield (σ_y) and maximum strength (σ_{max}). In contrast to experimental values found in the present study, Son et al. [14] establish that addition of graphite decreases the compressive strength of the composites, due possibly to a low contact area between matrix powders. Fogagnolo et al. [18] mention that reinforcement clusters, cracks in the reinforcement surface or poor bounding between matrix and reinforcements can also deteriorate the composite strength. High concentration composites (1.0%-CuMG) present a modest performance probably due to the presence of free enforced

142

particles which segregate forming agglomerates and lower the final properties [19] or by a matrix saturation effect [14]. Figure 4b shows a comparison of σ_y and σ_{max} in synthesized composites. The increment in both properties is noticeable, Cu-MG concentration and milling intensity have an important effect on the mechanical performance of the composites. It indicates a synergic effect of metal graphite addition and milling intensity. The optimum point is obtained with 4h of milling and a low concentration (0.5%-CuMG) of reinforcement particles, like Esawi et al. [20] found. Besides, Table II shows that compacted-sintered composites present an irregular trend, some samples exhibit an increase of hardness, while others present strong cracking and property loss. In this manner the P2 sample displays a hardness peak of all the pure Al series, but this is lower than the corresponding values for composites with MG additions. There is normally an optimum milling time for each composition according to Table II, further milling or a change in MG concentration produces normally lower values of hardness. The best response is obtained with a low MG addition (0.5%) and 4h of milling intensity. Microstructure differences and packing efficiency for sintering between milled powders explain the variation in the final hardness of the sintered sample.

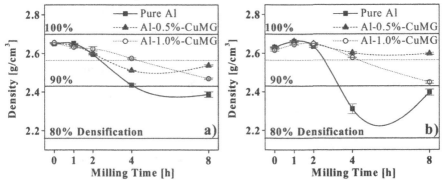

Figure 3. Density and densification percent curves of samples (a) before and (b) after sintering.

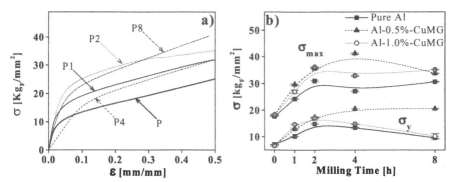

Figure 4. (a) Stress-Strain curves of un-doped Al samples. (b) Yield and maximum strength values found in sintered composites as a function of milling intensity and additive content.

CONCLUSIONS

Reinforcement particles are sub-micrometric and are homogeneously distributed leading to an important effect on the mechanical performance of the prepared composites. Additive concentration has an important effect on mechanical properties of composites and it has a synergic effect with milling intensity. Low concentration Al-CuMG composite with 4h of milling is the best option as a strengthening condition. Pre-milling process together with PM can increase the mechanical properties of Al-based composites prepared by this solid state route.

ACKNOWLEDGMENTS

This research is supported by CONACYT (Y46618). USA-Air force Office of Scientific Research, Latin America Initiative, Dr. Joan Fuller, contract # FA 9550/0 6/1/0524. Thanks to D. L. Gutierrez, A. H. Gutierrez, and E. T. Molle for technical assistance.

REFERENCES

1. J.M. Torralba, C.E. da Costa, F. Velasco. *J. Mat. Proc. Tech.* **133**, 203–206 (2003).
2. S.W. Ip, R. Sridhar, J.M. Toguri, T.F. Stephenson, A.E.M. Warner. *Mat. Sci. & Eng.* **A244**, 31–38 (1998).
3. H. Mayer, M. Papakyriacou. *Carbon* **44**, 1801–1807 (2006).
4. V. Amigo, J. L. Ortiz and M. D. Salvador. *Scripta Mater.* **42**, 383–388 (2000).
5. F. Akhlaghi, S.A. Pelaseyyed. *Mat. Sci. & Eng.* **A385**, 258–266 (2004).
6. C. Zubizarreta, S. Giménez, J.M. Martín, I. Iturriza. *J. Alloys & Comp.* **467**, 191–201 (2009).
7. J.B. Fogagnolo, F. Velasco, M. H. Robert, J.M. Torralba. *Mat. Sci. & Eng.* **A342**, 131-143 (2003)
8. M. Adamiaka, J.B. Fogagnolo, E.M. Ruiz-Navas, L.A. Dobrzañski, J.M. Torralba. *J. Mat. Proc. Tech.* **155–156**, 2002–2006 (2004).
9. D. Casellas, A. Beltran, J.M. Prado, A. Larson, A. Romero. *Wear* **257**, 730–739 (2004).
10. Hailong Wang, Rui Zhang, Xing Hu, Chang-An Wang, Yong Huang. *J. Mat Pross. Tech.* **197**, 43–48 (2008).
10. Naiqin Zhao, Philip Nash, Xianjin Yang. *J. Mat. Pros. Tech.* **170**, 586–592 (2005).
11. D. Oleszak, V.K. Portnoy and H. Matyja, *J. Met. & Nanocrys. Mats.* **2-6**, 345-350 (1999).
12. Z. Razavi Hesabi, A. Simchi, S.M. Seyed Reihani. *Mat. Sci. and Eng.* A **428**,159–168. (2006)
13. H.T. Son, T.S. Kim, C. Suryanarayana,, B.S. Chun. *Mat. Sci. & Eng.* **A348**, 163-169 (2003).
14. B.P. Neville, A. Rabiei. *Mat. & Design* **29**, 388–396 (2008).
15. Z. Razavi Hesabi, H.R. Hafizpour, A. Simchi. *Mat. Sci. & Eng.* **A454–455**, 89–98 (2007).
16. H. Abdoli, E. Salahi, H. Farnoush, K. Pourazrang. *J. Alloys & Comp.* **461**, 166–172 (2008).
17. J.B. Fogagnolo, M.H. Robert, J.M. Torralba. *Mat. Sci. & Eng.* **A426**, 85–94 (2006).
18. E.M. Ruiz-Navas, J.B. Fogagnolo, F. Velasco, J.M. Ruiz-Prieto, L. Froyen. *Composites* **A37**, 2114–2120 (2006).
19. Amal M.K. Esawi, Mostafa A. El Borady. *Comp. Sci. & Tech.* **68**, 486–492 (2008).

Mater. Res. Soc. Symp. Proc. Vol. 1243 © 2010 Materials Research Society

Microstructural Characterization of Multi-Component Systems Produced by Mechanical Alloying

R. Pérez-Bustamante[1], C.D. Gómez-Esparza[1], F. Pérez-Bustamante[2], I. Estrada-Guel [1], J.G. Cabañas-Moreno[3], J.M. Herrera-Ramírez[1], R. Martínez-Sánchez[1]

[1] Centro de Investigación en Materiales Avanzados (CIMAV), Laboratorio Nacional de Nanotecnología, Miguel de Cervantes 120, 31109 Chihuahua, Chih., México.
[2] Instituto Tecnológico de Chihuahua (ITCH), Av. Tecnológico 2909, 31310 Chihuahua, Chih., México.
[3] Instituto Politécnico Nacional – CNMN, UPALM, 07338 México, D.F., México.

ABSTRACT

A series of binary to hexanary alloys (Ni, Co, Mo, Al, Fe, Cu) are produced by mechanical alloying. Formation of an FCC solid solution is observed in the binary system. For ternary to quinary systems the presence of an amorphous phase and a BCC solid solution is identified, and for the hexanary system a combination of BCC and FCC solid solutions is detected. There is a very small change in the lattice parameter of Mo, reflecting the limited solid solubility of other element in this structure. However, Mo induces the fast amorphization of other elements and the reduction of crystallite size.

INTRODUCTION

Multi-component systems are characterized by a high entropy and ability to form amorphous phases [1-5]. High entropy alloys (HEAs) are a new era of materials that consist of various major alloying elements; the system selected in each case is different, and each additional element changes the behavior of the final alloy. The HEAs are quite simple to analyze and control because they tend to form simple solid solution phases, mainly of FCC and BCC structures. They have numerous beneficial mechanical, magnetic, and electrochemical characteristics [6]. These systems can be processed by different routes, conventional casting, thin film deposition and milling process. The mechanical alloying (MA) process has been widely recognized as an alternative route for the formation of nanocrystalline materials, with unusual properties [7, 8]. In this investigation, a multi-component system formed by Ni-Co-Mo-Al-Fe-Cu is studied from the binary to hexanary alloy; the effect of each element and milling time is reported and discussed.

EXPERIMENTAL PROCEDURE

Ni, Co, Mo, Al, Fe and Cu powders with purity higher than 99.5% and particle size of - 325 mesh are mechanically alloyed from binary Ni-Co to hexanary Ni-Co-Mo-Al-Fe-Cu systems in equiatomic ratio. Table 1 gives the nominal composition used in each system (identified as A, B, C, D, E series). The milling process is carried out from 0 to 30 h in a high energy shaker ball mill (SPEX-8000M). Hardened steel vials and balls are used as milling media. The milling ball-to-powder weight ratio is set at ~5:1. In order to avoid excessive welding of powders, methanol is added to the powders to act as a process control agent (PCA). The milled products are characterized by scanning electron microscopy (SEM) in a JEOL JSM-7401F microscope supplied with an energy dispersive spectrometer (EDS), operated at 5 kV and 20 µA. They are also analyzed by X-ray diffraction (XRD) in a Siemens D5000 diffractometer with Cu Kα radiation (λ=1.5406 Å) and operated at 35 kV and 25 mA in the 2θ range of 20-110°. The step and acquisition time are 0.2° and 5 s, respectively.

Table 1. Nominal* and experimental chemical compositions of different systems studied (at. %).

System	Milling time (h)	Ni	Co	Mo	Al	Fe	Cu
	0	*50.0	*50.0				
A	10	50.9	49.1				
	20	51.5	48.5				
	30	51.0	49.0				
	0	*33.3	*33.3	*33.4			
B	10	36.9	36.0	27.1			
	20	37.6	34.6	27.8			
	30	37.5	35.0	27.5			
	0	*25.0	*25.0	*25.0	*25.0		
C	10	29.1	28.9	21.7	20.3		
	20	29.3	26.6	22.1	22.0		
	30	29.1	26.8	22.0	22.1		
	0	*20.0	*20.0	*20.0	*20.0	*20.0	
D	10	21.0	22.6	17.1	17.6	21.7	
	20	21.4	21.7	17.3	17.7	21.9	
	30	21.6	22.0	17.3	17.5	21.6	
	0	*16.6	*16.6	*16.7	*16.7	*16.7	*16.7
E	10	17.5	18.2	14.2	15.2	16.8	18.1
	20	16.1	18.3	14.6	16.4	16.9	17.7
	30	17.1	18.0	14.3	15.9	16.9	17.8

RESULTS AND DISCUSSION

Figure 1 presents the XRD spectra from binary to hexanary systems mechanically alloyed for different times. In general, elemental characteristic peaks disappear as the milling time increases while peaks corresponding to the crystallization of a single solid solution appear. For all systems, this solid solution appears from 10 h of milling time. It is important to mention that for initial powders (0 h of milling) Co has an HCP structure, which transforms to an FCC structure after the milling process due to particle size reduction and the accumulation of structural defects, as reported elsewhere [9, 10]. Thus, in figure 1a, peaks from Co HCP disappear after 10 h of milling. Ni peaks (FCC structure) are broadened and shortened as the milling time is increased; additionally the position of the Ni diffraction lines is shifted to lower angles after 10 h of milling. This displacement of the peaks remains for all milling times and it is assumed to be an indication of the formation of a FCC solid solution by MA.

The XRD patterns as a function of milling time for the NiCoMo system are presented in figure 1b. Mo characteristic reflections decrease in intensity and become broader, but they are detected even after 30 h of milling, suggesting that this element is not dissolved in Ni or Co. By comparing figures 1a and 1b, an apparent effect of Mo on the amorphous phase formation is observed; for example, in figure 1a the two first peaks of Ni –in solid solution with Co– remain even after 30 h of milling, whereas in figure 1b Ni and Co reflections become too broad or disappear after 10 h of milling.

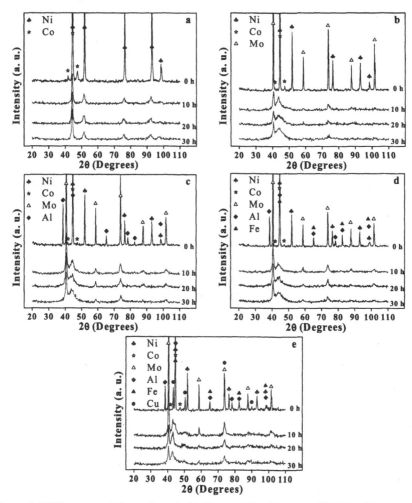

Figure 1. XRD spectra of the equiatomic (a) binary NiCo, (b) ternary NiCoMo,(c) quaternary NiCoMoAl, (d) quinary NiCoMoAlFe, and (e) hexanary NiCoMoAlFeCu systems as a function of milling time.

147

Figure 1c presents XRD patterns of the NiCoMoAl alloy. In this case, 10 h of milling considerably reduce the intensity of the diffraction peaks corresponding to Ni, Co or Al. Apparently 10 and 20 h of milling induce the formation of an FCC solid solution, but a longer milling time produces an amorphous phase. Nevertheless, the XRD patterns always show Mo reflections regardless of the milling time. In this alloying system, it is appreciated a small variation of Mo peaks to higher 2θ values as the milling time increases; this denote the incipient formation of a BCC solid solution.

In the case of the NiCoMoAlFe alloy, the XRD patterns are shown in figure 1d. There is a similar behavior as compared to the quaternary system. The reflections belonging to Co, Al and Fe disappear. Apparently, an amorphous phase formation is produced at long milling times while the reflections of Mo are still retained in the XRD pattern. However, in a similar case as the quaternary system, Mo reflections are shifted to higher angles as the milling time is increased, denoting the formation of a BCC solid solution.

For the NiCoMoAlFeCu system (Fig. 1e), there is no evidence of Co, Al, Fe and Cu peaks even after 10 h of milling. Milling produces only short and broad peaks near the original Mo peak positions and an apparent FCC solid solution. According to the peaks position, this FCC solid solution presents higher lattice parameter than that found in the binary system (Ni-Co system, Fig. 1a). Mo characteristic reflections present shortening, broadening and shifting to higher angles denoting crystal refining and the formation of a BCC solid solution.

Crystallite size and lattice strain of the alloys have been calculated from the diffraction peaks at 2θ positions near the first Ni and Mo peaks by using an algorithm included in the X'Pert Data Viewer software, based in the Scherrer formule; instrumental correction has been performed employing a silicon standard. Figure 2 shows the evolution of these parameters as a function of the milling time; each point corresponds to the average of 5 measurements and the error bars indicate the standard deviation. For all the series, a decrease in crystallite size and an increase in lattice strain are found as the milling time increases. This is attributed to the mechanical deformation introduced in the powders. Severe plastic deformation can lead to variation in crystallite size and the accumulation of internal stresses. Due to the precision achieved in our evaluations (5-15 Å), difference in crystallite size between samples B, C, D and E (Fig. 2a) could be negligible.

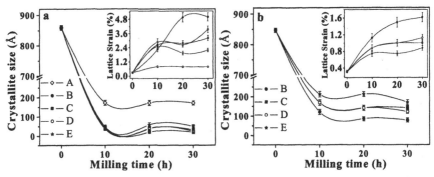

Figure 2. Crystallite size and lattice strain versus milling time for (a) first Ni-type and (b) first Mo-type reflections. A-E indicate the different investigated multi-component alloys.

Figure 3 shows the powder cross section microstructure of quaternary and hexanary alloys after 10 and 30 h of milling. For lower milling times a quasi-lamellar structure is observed, which is typical in the early stages of MA for ductile components [7]; during the SEM observations at higher magnifications, a bright phase is observed, which is identified as a Mo-rich phase by means of EDS microanalyses (Figs. 3a and 3c). With further milling, a more uniform microstructure (Figs. 3b and 3d) is observed, but some bright particles are still noted. These results are in accordance with the XRD analysis, where Mo reflections are detected in all systems containing this element (Figs. 1b-1e). General microanalyses by EDS show that the chemical composition of all systems is homogeneous after longer milling times (Table 1); oxygen signal is not detected, indicating that the oxidation during the milling process is very low.

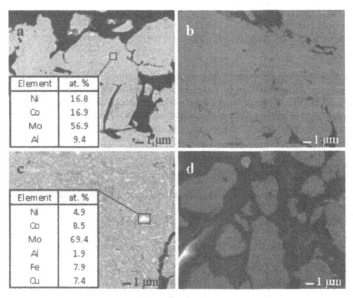

Figure 3. Powder cross section backscattered electron images of the quaternary NiCoMoAl alloy after (a) 10 h and (b) 30 h and the hexanary NiCoMoAlFeCu alloy after (c) 10 h and (d) 30 h of milling.

From the above results, a different answer is observed in all the alloying systems studied. Binary NiCo system shows the preference to form FCC solid solution. Ternary NiCoMo system shows a sequence Crystalline → BCC + FCC solid solution → BCC + Amorphous. Quaternary NiCoMoAl and quinary NiCoMoAlFe systems show the sequence Crystalline → BCC solid solution + Amorphous. Finally, the hexanary NiCoMoAlFeCu system shows the preference to form a BCC + FCC solid solution. For the binary system, despite the fact that for initial powder Co presents an hexagonal structure, the final phase formed is an FCC solid solution, whose reflections correspond with Ni peaks positions. From ternary to quinary systems, the apparent amorphous phase formation after longer milling time is present. For ternary system, Mo shows

149

an apparent inertness to combine with Ni or Co; Mo peaks shifting is not observed. However additions of Al (quaternary system) and Al-Fe (quinary system) have an effect in the Mo peaks positions, they are shifted to higher values. Mo effect on the amorphous phase formation could be attributed to the i) atomic size factor, as reported by Miedema [11], ii) difference in crystal structure, and iii) difference in melting point. In the case of the FCC solid solution peak measurements (Fig. 2a), the presence of Mo in systems B, C, D and E induces the reduction of the crystallite size, giving rise to the formation of an apparent amorphous phase, and the increment of the lattice strain, in comparison to the NiCo system (series A). For the BCC solid solution peak measurements (Fig. 2b), the reduction of crystallite size and lattice strain increase are due just to the milling effect [7]. In both FCC and BCC solid solution peak measurements, higher values in lattice strain can be seen for the alloy system B. From figure 2b, quaternary to hexanary systems have a larger BCC solid solution crystallite size than the ternary (NiCoMo) system. The crystal size refinement and micro-tensions increment are normal answers in powder mechanically alloyed.

CONCLUSIONS

The binary to hexanary equiatomic high entropy alloys in Ni-Co-Mo-Al-Fe-Cu system have been successfully synthesized by mechanical alloying in a high energy ball mill. Formation of an FCC solid solution structure after longer milling time was observed in binary and hexanary systems. After 10 h of milling, it was observed that the Mo presence improves the amorphous phase formation in ternary and quaternary systems. Mo presence favored the BCC solid solution structure formation.

ACKNOWLEDGEMENTS

Thanks to W. Antúnez-Flores and E. Torres-Moye for their valuable technical assistance.

REFERENCES

1. Y. Zhang, Y.J. Zhou, X. Hui, M. Wang and G.L. Chen, Science in China Series G: Physics, Mechanics & Astronomy **51**-4 (2008) 427-437.
2. Y.L. Chen, Y.H. Hu, C.W. Tsai, C.A. Hsieh, S.W. Kao, J.W. Yeh, T.S. Chin, S.K. Chen, J. Alloys Compd. **477** (2009) 696-705.
3. H. Xie, J. Lin, Y. Li, P.D. Hodgson, C. Wen, Mat. Sci. & Eng. A **459** (2007) 35-39.
4. S. Varalakshmi, M. Kamaraj, B.S. Murty, J. Alloys Compd. **460** (2008) 253-257.
5. T.K. Chen, T.T. Shun, J.W. Yeh, M.S. Wong, Surf. Coat. Technol. **188-189** (2004) 193-200.
6. C.P. Lee, Y.Y. Chen, C.Y. Hsu, J.W. Yeh and H.C. Shih, J. Electrochem. Soc. **154** (2007) C424-C430.
7. C. Suryanarayana, Progress in Materials Science **46** (2001) 1-184.
8. L. Lü and M.O. Lai, Mechanical Alloying, Kluwer Academic Publishers, Boston, 1998
9. S. Ram, Mat. Sci. & Eng. A **304-306** (2001) 923-927.
10. J.Y. Huang, Y.K. Wu and H.Q.Ye, Acta mater. **44** (1996) 1201-1209.
11. F.R. de Boer, R. Boom, W.C.M. Mattens, A.R. Miedema, A.K. Niessen, *Cohesion in Metals, 2nd printing*, North-Holland, New York, 1989.

Mater. Res. Soc. Symp. Proc. Vol. 1243 © 2010 Materials Research Society

On the Solidification and Feeding of a Ductile Iron Casting

Eudoxio A. Ramos Gómez[1], Marco A. Ramírez-Argáez[2], Carlos González-Rivera[2].
[1] PceIM, Universidad Nacional Autónoma de México, Mexico City, Mexico,
[2] Facultad de Química, Universidad Nacional Autónoma de México, Mexico City, Mexico

ABSTRACT

Solidification of a simple casting made of ductile iron is mathematically modeled in this work. The model is able to numerically simulate the cooling rate and solidification of the whole casting system composed by a cubic piece, a blind riser connected through a rectangular neck, and immersed in a green sand mold. The center of the neck acts as a valve that allows the flow of liquid metal between the casting and the riser based on the feeding technique known as Pressure Control Risering (PCR). The developed model couples the energy conservation equation and the solidification kinetics of ductile iron, through the statement of proper nucleation and growth laws. This model is satisfactorily validated by comparing the thermal histories predictions with experimental cooling curves obtained in the foundry laboratory for the same casting. According to a process analysis developed in this work, the pouring temperature is the variable that affects the most the solidification and the feeding behavior, since it increases significantly the solidification times in all regions of the casting system.

INTRODUCTION

Porosity due to contractions is still one of the most frequent rejecting causes of ductile iron sand castings. Despite these phenomena are not new, mechanisms involved in the appearance of these defects in ductile iron differ significantly from other metals. From experimental evidence, it is currently accepted that ductile iron suffers an expansion during solidification that builds up a pressure high enough to plastically deform the sand mold to yield swollen castings that contain contraction defects [1, 2]. The implementation of a risering method is a way to avoid the presence of these defects since it controls the degree of pressure due to expansion during solidification of the casting. The risering technique called Pressure Control Risering (PCR) controls pressure in a way that the liquid iron is always under a positive pressure relative to the atmosphere, but never reaches high values of pressure that would plastically deform the green sand mould [3].

THEORY

The casting studied in this work consists of a simple iron cube (casting-neck-feeder) contained in a green sand mould that transfers heat to the surroundings by free natural convection (see Figure 1). It is assumed that initially the entire casting plus the feeder are filled with liquid iron of eutectic composition being all initially at the pouring temperature. It is also assumed that conduction controls heat transfer and thus convection is neglected in liquid iron, volume changes are not considered in the calculations. The system is assumed to be continuous and therefore the gap formed between the casting and the sand mould is not taken into account in the present model. The energy released is only due to the eutectic solidification of ductile iron, the remaining heat transformations in the solid state are neglected.

Cooling and solidification of the casting are simulated by a model developed in this work that couples the energy conservation equation with a micro model that describes the solidification kinetics of the eutectic based on nucleation and growth models. Conservation of energy (Ec. (1)) represents an energy balance that considers only the net heat flow of conduction and the source term associated to the heat of eutectic solidification equal to the energy change with time.

$$\nabla \cdot k\nabla T + S_T = \frac{\partial \rho C_P T}{\partial t} \tag{1}$$

Where ∇ is the nabla operator, T is temperature, ρ, k, and C_p are the density, the thermal conductivity and the specific heat of the media respectively, while t is the time.

The energy source term, S_T (Ec. (2)) is directly proportional to the solid fraction evolution, $\partial f_s / \partial t$. The solid fraction growth rate term is important since it allows the coupling of the micro model of solidifications kinetic and the macro model of heat transfer.

$$S_T = \lambda_f \frac{\partial f_s}{\partial t} \tag{2}$$

The proportionality constant of Ec. (2), λ_f, is the latent heat of fusion of ductile iron. The solid fraction evolution is computed from the micro model "graphite nodules – austenite shell", based on the growth law developed by Wetterfall et al [4]. This micro model assumes at the beginning of solidification carbon precipitates in the shape of spherical nodules covered by an austenite shell. This spherical composite grain will further grow by diffusion of carbon through the austenite phase.

The sand mould is exposed to natural convection of air at 291 K on five of its faces as illustrated in Figure 1(a) where also the casting is seen along with the feeder and the neck connecting them. The heat transfer coefficient is set to a constant value but its value is different for a vertical mould wall than for a horizontal surface. In this work an average value of 50 W/m²*K is set, corresponding to natural free air convection [5]. Additionally, proper thermal conductivities and specific heats values are needed as a function of temperature for the sand mould, solid and liquid iron respectively [6] due to the presence of different phases in the iron and the wide range of temperatures involved during solidification. For iron in the mushy zone, properties are set as a weighted average based on the instantaneous solid fraction value.

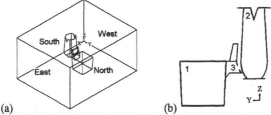

(a)   (b)

Figure 1.- Location of casting, riser and boundaries of the green sand casting system studied; (a)mold/metal system, (b)metal part of the system

Figure 1(b) shows the locations of the thermocouples used during thermal validation of the model, with the numbers 1 for the casting, 2 for the riser and 3 for the neck. Table I shows the more relevant parameters of the model.

152

Table I. Parameters of the model

Properties	Liquid Metal	Solid Metal	Sand Mould
Thermal Conductivity (W/m*K)	1.1164*T-1597.8	-30.889Ln(T)+243.64	8E-7*T²-6E-4*T +.6611
Specific heat (J/kg*K)	3E-4*T²-0.2382*T+598.96	3E-4*T²-0.2382*T+598.96	-3E-4*T²-0.8038*T+576.82
Density (kg/m³)	7000	7000	1700
Initial Temperature (K)	Pouring Temperature	------------------------------	298

A numerical solution has been implemented for the partial differential equation representing the conductive heat transfer problem with solidification, through the commercial computational fluid dynamics (CFD) software PHOENICS™. An orthogonal grid mesh with more than 75,000 nodes is built and the metallic bodies are shaped by using a Body Fitted Coordinate (BFC) module included in the CFD code. A time step of 1 second is selected. Additionally, Fortran subroutines have been written and coupled to PHOENICS™ to include the micro model of solidification.

The model is validated by comparing experimentally obtained versus predicted cooling curves. Thermocouples have been positioned in the geometric centre of the neck, and in selected locations of the cube and riser (see Figure 1(b)). The validation experiment is made by triplicate using an eutectic ductile iron with a chemical composition in weight percent of 3.6 % C and 2.1 % Si. Figure 2 shows experimental and predicted cooling curves at three different locations (Points 1, 2 and 3 in Figure 1(b)) during stages of cooling of the liquid, solidification and cooling of the solid iron. These stages are represented as the L, L+S and S regions on Figures 2(a) to (c). It is clear that, in general a very good agreement is found between the model predictions and experimental results in terms of liquid cooling rate and solidification time. Perhaps the only difference is a slower cooling rate in the solid state predicted by the model relative to the experimental cooling curves. This good agreement between computed and experimentally determined cooling curves suggest that the micro macro model developed is successfully validated and that it can be able to predict heat transfer and solidification kinetics of eutectic ductile iron in the system under study.

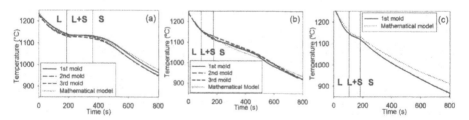

Figure 2. Predicted and experimental cooling curves at (a) neck, (b) casting and (c) feeder.

RESULTS AND DISCUSSION

A process analysis is performed to study the effect of the main process variables on the cooling rates, solidification time and microstructure characteristics of the casting. Figure 3 shows predicted cooling curves at the center of the neck (Point 3 in Figure 1(b)) and varying either (a) the number of nodules/m^3, (b) the neck transversal section, or (c) the pouring temperature. The number of nodules does not affect the cooling curves. The pouring temperature has a major effect on the cooling curves followed by the transversal section of the neck. Increasing pouring temperature (Figure 3(c)) produces a delay in the beginning of solidification since the system requires more time to reach the eutectic temperature. Additionally, there is an increment in the amount of energy transferred from the riser and casting to the neck, which results in a higher solidification time. On the other hand, an increment in the section of the neck is equivalent to increase the dissipated energy, which results in a lower cooling rate (Figure 3(b)) and a higher solidification time.

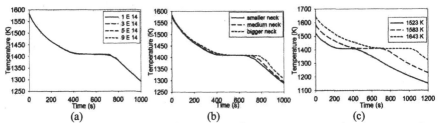

(a)   (b)   (c)

Figure 3. Effect of the process variables on the cooling curves at center of neck represented as the point 3 in Figure 1(a)): (a) Number of nuclei; (b) Size of the neck; (c) Pouring temperature.

The effect of the different variables on the solidification kinetics can be followed more clearly by analyzing the variation of the solid fraction as a function of time. This can be done by obtaining the solidification sequence, that is the time at which solidification ends in different parts of the system under study. In order to get this sequence properly, the entire casting is divided in several parts: cylinder, semi sphere, cube (casting), neck and center of the neck. The central part of the neck (which we call valve) is the most important part of the system because it regulates the liquid exchange between casting and riser and it is assumed to represent approximately 30% of the neck transversal section. Once its solid fraction reaches a critical value, the valve will close avoiding any transfer of liquid metal from the riser to the cube or viceversa.

The time of end of solidification is shown in Figure 4 as a function of the process variables. Model results show that the effect of the number of nuclei on this parameter is negligible under conditions considered in this work. On the other hand it is seen in Figure 4(b) that increasing pouring temperature increases the time to complete solidification in all zones of the system. Neck size variation has only a local effect on the valve and the neck retarding the end of solidification (Figure 4c).

Taking into account that the model outcome shows, see Figures 3 and 4, that pouring temperature is the most important process variable, its effect on the solid fraction evolution in the cube and valve is presented in Figure 5.

In Figure 5 it is clear that as the pouring temperature increases the beginning and the end of solidification are retarded in the entire system. Solidification time, defined as the time difference between the beginning and the end of solidification is increased when pouring temperature increases and this effect can be seen on Figure 5 from the slope changes of solid fraction curve of the cube, while for the valve, this effect is observed as a slower start of solidification (see the arrows in Figure 5). Solid fraction evolution rate is faster in the valve than in the cube since the former implies solidification of less amount of metal. The most important feature of Figure 5 is the time at which both curves intersect. As pouring temperature decreases, solid fraction curves intersect at lower values of solid fraction or shorter times (Figure 5(a)), which means that the valve of the neck will close at earlier stages of the process. Intersection between solid fraction curves is important because this determines the amount of liquid metal in the casting when the valve closes which in turn determines if the casting can be obtained free of contraction defects, or if it will be rejected due to the presence of a swollen casting or the presence of contraction porosity in the cube.

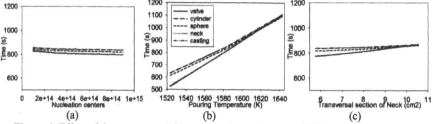

| (a) | (b) | (c) |

Figure 4. Effect of the process variables on the time of end of solidification: (a) Number of nuclei; (b) Pouring temperature; (c) Size of the neck.

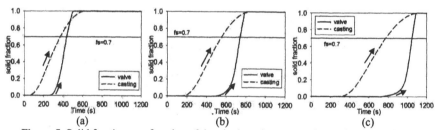

| (a) | (b) | (c) |

Figure 5. Solid fraction as a function of time in the valve and casting regions for different pouring temperatures (a) 1523 K, (b) 1583 K and (c) 1643 K.

It is well known that as soon as the solid fraction reaches a value of 0.7, the flow of the remnant liquid stops and remains trapped between interdendritic arms [7]. Such a value is important and will be used here, since the present model has not yet been extended to predict the casting solid fraction for the valve to close. According to this, the following analysis is presented. Figure 6 shows the effect of process variables on the solidification with a closed valve (solidification fraction > 0.7). As can be seen, all process variables have a positive effect on the solidification kinetics i.e., the solid fraction increases for all zones of the system when the valve closes. Again, pouring temperature is the most significant variable (Figure 6b). For the lowest pouring temperature a 0.82 value of solid fraction is obtained in the cube when valve closes indicating the possibility that pressure from the remnant liquid may plastically deform the mould leading to a swollen casting. In contrast, for the highest pouring temperature, the remnant liquid is so little when valve closes that it is feasible that a secondary contraction may not be compensated by carbon precipitation leading to a casting potentially rejected by porosity defects.

To determine if an early or a late solidification of the neck may lead a sound or unsound casting it is necessary to include in the model possible volume changes produced during solidification as a function of the chemical composition and temperature.

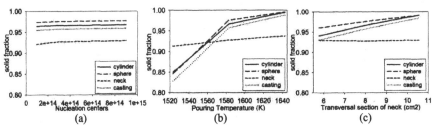

(a)   (b)   (c)

Figure 6. Effect of process variables on the solid fraction when valve has closed: (a) Number of nuclei; (b) Pouring temperature; (c) Size of the neck.

CONCLUSIONS

A mathematical model that simulates heat transfer and solidification kinetics is developed and validated. The model is used to perform a process analysis on a simple casting system including a riser, a neck and a sand mould, exploring the effect of number of nucleation sites, pouring temperature and transversal section of the neck on the solidification kinetics of the system. Pouring temperature is the most important variable affecting solidification kinetics of the system and increasing this variable solidification kinetics of the system is retarded. Changes in pouring temperature affect solidification time of the neck with respect to the end of solidification of the casting, which could be relevant to the correct performance of the feeding system. An increase in the transversal section of the neck produces a local effect affecting mainly the neck zone, retarding its solidification kinetics which also could be useful to control the feeding ability of the system. The effect of number of nuclei is negligible compared to the effects produced by changes in pouring temperature and neck size.

REFERENCES

1. K. S. Lee, M. Kayama, *IMONO*, **30**, 11 (1976)
2. G. Nadori in *International Foundry Congress* Paper, **15**, (1978).
3. G. A. Corlett, *AFS Transactions* **90**, 173 (1983).
4. Wetterfall, H. Fredriksson and M. Hillert, *Journal of the Iron and Steel Institute*, 323 (1972).
5. Geiger and Poirier, *Transport Phenomena in Materials Processing*, (TMS, 1994) pp. 258-261.
6. R. D. Pehlke, A. Jeyarajan and H. Wada: Summary of thermal properties for casting alloys and mould materials NITS/PB83/211003, University of Michigan, Ann Argor, MI, 1982
7. John Campbell, *Castings*, 2nd ed. (2003, Ed. Butterworth), pp. 218-244.

Mater. Res. Soc. Symp. Proc. Vol. 1243 © 2010 Materials Research Society

Physical Modeling of Gas Jet-Liquid Free Surface in Steelmaking Processes.

J. Solórzano-López[1], R. Zenit[2], C. González-Rivera[1] and M. A. Ramírez-Argáez[1].
[1]Facultad de Química, UNAM, Av. Universidad 3000, Coyoacán, 04510, México, D.F.
[2]Instituto de Investigaciones en Materiales, UNAM, Av. Universidad 3000, Coyoacán, 04510, México, D.F.

ABSTRACT

Gas jets play a key role in several steelmaking processes as in the Basic Oxygen Furnace (BOF) or in the Electric Arc Furnace (EAF). They improve heat, mass and momentum transfer in the liquid bath, improve mixing of chemical species and govern the formation of foaming slag in EAF. In this work experimental measurements are performed to determine the dimensions of the cavity formed at the liquid free surface when a gas jet impinges on it as well as liquid velocity vector maps measured in the zone affected by the gas jet. Cavities are measured using a high speed camera while the vector maps are determined using a Particle Image Velocimetry (PIV) technique. Both velocities and cavities are determined as a function of the main process variables: gas flow rate, distance from the nozzle to the free surface and lance angle. Cavity dimensions (depth and diameter) are statistically treated as a function of the process variables and also as a function of the adequate dimensionless numbers that govern these phenomena. It is found that Froude number and Weber number control the depression geometry.

INTRODUCTION

Oxygen jets are widely used in steelmaking processes. These jets play important roles in controlling chemical reaction kinetics, forming foaming slags, bath mixing and splashing phenomena since they exchange momentum, heat and mass with the steel bath [1,2]. It is difficult to measure these parameters in actual process conditions [3]. However, water physical modeling is a useful tool to perform studies in a laboratory scale at very low cost and safe working conditions[3-7]. The mathematical models are also useful [8-10] for process simulation at low cost. In this work, experimental measurements of the geometry of the cavity formed by the impingement of a gas jet on a water free surface are reported together with velocity profiles of the liquid bath driven by the momentum transfer of the gas jet.

EXPERIMENT

A cylindrical (height=0.3 m, internal diameter=0.2 m) water model is constructed using transparent acrylic. The model is enclosed in a rectangular tank to reduce image distortion. A cylindrical lance of 0.012 m internal diameter is made of copper and coupled to a cylindrical nozzle. An air compressor and a flow meter are used to regulate the gas flow rate. Measurements of the dimensions of the cavity are made using a high speed camera (HSC) at 500 frames per second with an exposure time of 1/500 s during 2 s. Photographs included a length scale to allow measurement of width and depth of the cavity. These measurements are processed statistically by applying multivariable linear regressions.

A PIV system (Dantec Dynamics) and analysis software (Flow Manager 4.5) are used to obtain the liquid velocity maps in the bath region affected by the jet. The water is seeded with glass spheres which acted as tracers. These are needed to perform the cross correlations required to measure the velocity profiles. Images are properly masked to discard zones near the point of impingement of the gas jet. Time averaged velocity profiles are finally obtained for each set of experimental conditions and the experiments are performed at least three times to achieve statistical significance. The variables considered in this study are lance angle (60°-90°), gas flow rate in the jet (60-120 l/min) and lance height (20-100 mm). A new variable called distance from nozzle tip to bath, z, is determined by trigonometry in terms of lance angle and height (see Figure 1).

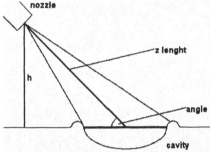

Figure 1. Distance from nozzle tip to bath, z, obtained from the lance height, h, and the lance angle.

DISCUSSION

Cavity dimensions
By measuring the dimensions of the depth and width of the cavity (see Figure 2), it is found that some variables affected the cavity dimensions more than others. For instance, it is found that the width of the cavity is directly proportional to the gas flow rate while the z distance does not have a significant effect on the cavity width.

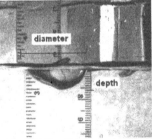

Figure 2. Typical image of the cavity formed on the free surface by the gas jet impingement. The diameter and depth are indicated.

On the other hand, the depth of the cavity is inversely proportional to the distance z and directly proportional to the gas flow rate. In general, it is observed that the size of the cavity is smallest when the impinging jet formed at low flow rates comes from a large z. At a gas flow rate of 120 l/min the depth increases. Vertical blowing (90°) decreases the width of the cavity but the depth increases since all the jet momentum is perpendicular to the bath surface and the jet penetration increases (see Figure 3).

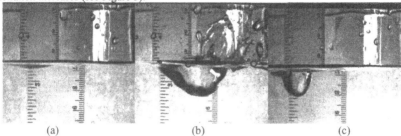

(a)   (b)   (c)

Figure 3. Effect of jet characteristics on cavity geometry under different experimental conditions: a) gas flow rate 80 l/min, lance angle of 60° and lance height (h) 100 mm, b) gas flow rate 120 l/min, lance angle 60° and lance height (h) 50 mm, c) gas flow rate 80 l/min, lance angle 90° and lance height (h) 50 mm.

Empirical correlations

In order to provide a more generalized correlation of the geometry of the cavity with the process variables, an statistical analysis is performed based on the main dimensionless groups that govern the geometry of the cavity, i.e. Froude (Fr), Weber (We), and Reynolds (Re) numbers, after a dimensionless analysis of the fluidynamics of the jet-bath system.

The analysis produced the following correlations of the dimensionless cavity width (a/z) and depth (Hj/z) with the Weber and Froude numbers:

$$\frac{a}{z} = 0.3 * We^{0.23} \tag{1}$$

$$\frac{Hj}{z} = 0.082 * \frac{We^{0.493}}{Fr^{0.171}} \tag{2}$$

The correlation coefficients are 0.975 and 0.995 for equations 1 and 2, respectively. It is interesting to note that cavity width is only a function of the Weber number; therefore this width depends on the ratio between inertial force of the gas jet and surface tension force. On the other hand, the cavity depth is a function of both Fr and We numbers, indicating that the depth of the cavity is controlled by a force balance between the inertial force of the jet, the buoyant force of the liquid and the surface tension force. These relationships, based on force balances to describe the free surface deformations, have been already reported by several authors [5, 11, 12, 13]. In the present system it appears that the We number plays a more important role that the Fr number. This observation indicates that the inertial forces of the jet must overcome first the surface tension forces and then the buoyant forces to produce the characteristic depth of the cavity.

Wakelin and Bradshaw have studied these phenomena[1]. The present experimental and calculated data (Figure 4) agrees well with the correlation proposed by these researchers:

$$\left(\frac{h}{z}\right)\left(\frac{h+z}{z}\right)^2 = \left(\frac{Jz}{g * \rho_l * z^3}\right) \tag{3}$$

159

where h is the depth of the cavity, z is the distance from the nozzle tip to the bath surface, J_z is the jet momentum transfer rate, g is the gravitational constant and ρ_l is the density of the liquid.

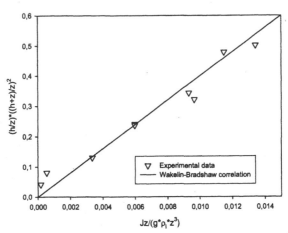

Figure 4.- Experimental results on cavity depth plotted according to the empirical correlation by Wakelin and Bradshaw.

Velocity profiles

Results of time averaged velocity profiles obtained by PIV in the liquid bath zone affected by the action of the air gas jet near the cavity are described next under different process conditions. In general, the trends indicate that velocities increase with the gas flow rate and decrease when the distance from the nozzle tip to the bath surface increases. The direction of motion of water determined by the gas flow direction which, in turn, is defined by the position of the lance (50 mm lance height and lance angle of 60°). The depth and width of the cavity as well as the jet penetration (amount of fluid affected by the jet momentum transfer) increase as the gas flow rate increases. Maximum bath velocities of 0.039 and 0.044 m/s are measured under the highest gas flow rates conditions of 100 and 120 l/min, respectively (Figure 5).

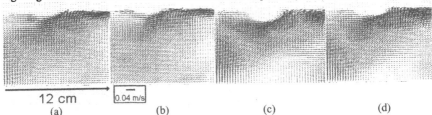

12 cm   0.04 m/s
    (a)   (b)   (c)   (d)

Figure 5.- Time averaged liquid velocity vector profiles obtained by PIV with different gas flow rates: a) 60 l/min, b) 80 l/min, c) 100 l/min and d) 120 l/min. The rest of variables remain constant (lance height 50mm and lance angle 60°).

160

The following results are obtained by varying the lance angle and keeping the rest of conditions constant (lance height of 50 mm and gas flow rate of 80 l/min). As can be see, as the lance angle deviates 90°, liquid velocity vectors show a higher horizontal component since the shear stress of the jet became higher as the lance is set more horizontally. The cavity depth decreases since the normal momentum or the impingement pressure of the jet decreases as the lance is placed more horizontally. A vertical lance (90°) under the same process conditions promotes the highest depression depth. A combination of a small distance between the lance and the free surface and a high gas flow rate produces a maximum cavity depth of 17.81 mm (Figure 6).

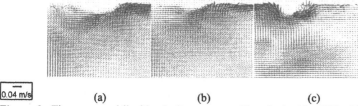

0.04 m/s   (a)   (b)   (c)

Figure 6.- Time averaged liquid velocity vector profiles obtained by PIV with different lance angles: a) 60°, b) 75°, and c) 90°. The rest of variables remain constant (lance height 50mm and gas flow rate 80 l/min).

At different lance heights (lance angle of 60° and gas flow rate 80 l/min) the liquid bath velocities and cavity sizes increase as distances from the nozzle to the bath decrease. This effect is due to the increase in jet momentum which increases the both the shear stress from the jet to the bath and the impingement pressure (Figure 7).

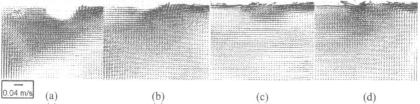

0.04 m/s   (a)   (b)   (c)   (d)

Figure 7.- Time averaged liquid velocity vector profiles obtained by PIV with different lance heights : a) 20 mm, b) 50 mm, c) 80 mm, and d) 100m. The rest of variables remain constant (lance angle 60° and gas flow rate of 80 l/min).

CONCLUSIONS

The results of the present investigation show that when a air jet impinges on a flat liquid surface the width of the cavity formed depends only on the Weber number. This result indicates that inertial forces are predominant over surface tension forces. The depth of the cavity depends more on the inertial forces than the surface tension and buoyant forces. Measured velocity profiles indicate that liquid velocity increases with air flow and when the nozzle distance from the free surface decreases.

161

ACKNOWLEDGMENTS

Authors thank to the CONACyT to provide financial support for this work through the Grant 60033 and through a doctoral scholarship. Additionally, authors would like to thank to Professor Adrian Amaro-Villeda for his support in the development of the experiments in this work.

REFERENCES

1. Wakelin, D. H. Ph. D. thesis, University of London, 1966.
2. Lee, M., Whitney, V., Molloy, N., Scandinavian Journal of Metallurgy, **30**, 2001, 330-336.
3. Sharma, S.K., Hlinka, J.W., Kern, D.W., Iron and Steelmaker, **4**, 1977, No. 7, 7-18.
4. Nordqist, A., Kumbhat, N., Jonsson, L., Jonssön, P., Steel Research International, **77**, 2006, No. 2, 82-90.
5. Koria, S.C., Lange, K.W., Steel Research, **58**, 1987, No. 9, 421-426.
6. Qian F., Muthasaran, R., Farouk, B., Metallurgical and Materials Transactions B, **27B**, 1996, No. 6, 911-920.
7. Subagyo, Brooks, G. A., Coley, K. S. and Irons, G. A., ISIJ International, **43**, 2003, No. 7, 983-989.
8. Tago Y., Higuchi Y., ISIJ International, **43**, 2003, 209-215.
9. Gu L., Irons G., 1999 Electric Furnace Conference Proceedings, 269-278.
10. Schwarz M. P., Fluid Flow Phenomena in Metals Processing, The Minerals, Metals & Materials Society, 1999, 171-178.
11. Memoli, F., Mapelli, C., Ravanelli, p., Corbella, M., ISIJ International, **44**, 2004, No. 8, 1342-1349.
12. Banks, R. B., Chandrasekhara, D. V., Journal of Fluid Mechanics, **15**, 1963, 13-34.
13. Mc Gee, P., Irons, G. A., 1999 Electric Furnace Conference Proceedings, 439-446.

Mater. Res. Soc. Symp. Proc. Vol. 1243 © 2010 Materials Research Society

Metallurgical Effects of Solution Heat Treatment Temperatures of Alloy Haynes™ HR-120™ for Land-Based Turbine Application

O.Covarrubias[1,2] and Osvaldo Elizarrarás[1]

[1]Frisa Aerospace SA de CV, Valentin G Rivero 200, Colonia Los Treviño, Santa Catarina, Nuevo León, México, 66150

[2]Facultad de Ingenieria Mecanica y Electrica, UANL, Ciudad Universitaria, San Nicolas, Nuevo León, México 66500.

ABSTRACT

Haynes™ HR-120™ alloy is a solid-solution-strengthened heat resistant alloy. The main characteristics of this alloy are strength at elevated temperature combined with resistance to carburizing and sulfidizing environments. Typical solution heat treatment for this alloy is usually performed above 1100°C. Solution heat treatment promotes non-desired precipitates to dissolve and, if deformation parameters are adequate, re-crystallization after forging procedures. It is reported that the solution temperature can also promote non-controlled grain coarsening. This investigation summarizes results on the effect of solution heat treatment on the microstructure of forgings when it is performed at 1000°C, 1050°C and above 1100°C. The experimental conditions resemble industrial environments. The obtained results include the alloy microstructural evolution by optical microscopy and Scanning Electron Microscopy (SEM) and the effect of these heat treatments on mechanical properties such as tensile, hardness and stress-rupture properties.

INTRODUCTION

Haynes™ HR-120™ alloy is a solid-solution-strengthened heat resistant alloy. The main characteristics of this alloy are strength at elevated temperature combined with resistance to carburizing and sulfidizing environments. Considering these properties, Haynes™ HR-120™ alloy is suitable for the production of components of land-based turbines, including rings. These rings can be manufactured by forging processes, where temperature and deformation ratios must be defined; considering heat treatment procedures, temperature and soaking are crucial for microstructural characteristics which promote required mechanical properties and microstructure [1]. The nominal chemical composition of alloy Haynes™ HR-120™ is given in Table 1 [2].

Table 1. Typical Chemical Composition for Haynes™ HR-120™, Wt%

Fe	Ni	Cr	Co	Mo	W	Cb	Mn	Al	C
33.0	37.0	25.0	3.0	2.5	2.5	0.7	0.7	0.1	0.05

Due to its mechanical ductility characteristics, this material can be readily formed on hot or cold conditions, making it suitable for the production of components as incinerators, muffles, retorts, recuperators and other components of land based turbines [3, 4]. Some of these components are produced by ring-rolling processes, are parameters like temperature, and deformation ratios have important effects in microstructural characteristics like grain size.

In general, after forming/rolling operations Haynes™ HR-120™ is exposed to solution heat treatments in the temperature range from 1175 to 1230°C, followed by a rapid cooling as rapid-air, polymer or water quench [5]. These procedures promote grain size homogenization and dissolution of non-desired precipitates. A typical solution heat treatment for this alloy is considered to be above 1100°C. Solution heat treatment promotes non-desired precipitates to dissolve and, if deformation parameters are adequate, re-crystallization after forging procedures. It is reported that a high solution temperature can also promote non-controlled grain coarsening [1]. Nevertheless the grain growth effect on mechanical properties such as hardness or tensile strength is considerably weak unless the application is extremely specific. Another general requirement to consider is the uniformity of microstructure, since duplex or bimodal grain size distributions, wide range or necklace conditions are hardly allowed. Thus in this research, the alloy Haynes™ HR-120™ has been exposed to different solution temperatures (1000°C and 1050°C) in order to evaluate their effect on mechanical properties and microstructural characteristics of seamless rings. Specific process parameters such as upsetting and ring-rolling conditions of evaluated rings are not covered in this report.

EXPERIMENTAL PROCEDURE

Several rings are produced with alloy Haynes™ HR-120™ considering industrial conditions; each ring has a weight of 73 Kg, and a geometry of 643mm OD x 510mm ID x 75 mm Height. Upsetting operations are performed by a hydraulic press of 3500 Ton. Ring-rolling is performed by a horizontal ring mill of 400 Ton. Selected parameters as forging temperatures and deformation ratios are set for the promotion of a desired "as forge" condition microstructure, where grain size is required as finer than ASTM 4.0 (89.8μm), and allowing coarse grains "as large as" (ALA) ASTM 2.0 (179.6μm).

Samples of such rings are extracted by segmentation and exposed to the solution heat treatment conditions listed in Table 2. The objectives for the considered solution heat treatments are (a) the promotion of required mechanical properties (as tensile, stress-rupture strength and hardness) and, (b) control on the uniformity and re-crystallization phenomenon, where grain size will be monitored to fulfill the above mentioned restrictions.

Table 2 Solution heat treatment conditions			
Condition	Temperature °C	Soaking time h	Cooling media
S1	1000	0.5	Water Quench
S2	1050	0.5	Water Quench
S3	>1100	0.5	Water Quench

Samples extracted from heat-treated segments are tested for tensile at room temperature per ASTM E8, hardness per ASTM E10 and smooth stress rupture (SSR) per ASTM E139. Optical microscopy is performed to evaluate grain size per ASTM E112 and microstructure uniformity per ASTM E1181. Specimens for microstructural evaluations are transversal segments, non-heat treated for "as forge" condition and heat treated for the three solution conditions. Scanning Electron Microscopy (SEM) is performed for detailed observation of microstructural characteristics when non-stressed end of representative tensile samples are

164

evaluated. Energy-dispersive X-ray spectroscopy (EDX) is performed for semi-quantitative composition determination of detected intergranular precipitates.

For stress-rupture testing, two conditions of load and temperature are evaluated: approx. 20.0 MPa at 982°C and approx. 50.0 MPa at 899°C. For all stress-rupture tests applied load is constant and when a determined time to rupture is reached, test is interrupted.

RESULTS AND DISCUSSION

Mechanical properties

The evaluation of mechanical properties promoted by the solution heat treatments is summarized in Table 3. It is found that despite heat treatment conditions, strength properties on tensile testing are only slightly affected: the three heat treatment conditions promote are practically the same. Yield strength (YS) and ultimate tensile strength (UTS) are similar for different samples exposed to three different solution heat-treatment conditions. More evident differences are related to elongation properties (%E), higher values are promoted when the alloy is exposed to solution above 1100°C (condition S3). As expected and considering tensile properties, the measurements of hardness (HBW) are similar for the three solution heat treatments. For stress-rupture evaluations (SSR), all specimens fulfill the requirements i.e., no single failure is reported. This behavior of the alloy Haynes™ HR-120™ demonstrates its consistency when exposed to the described conditions of load and temperature, despite previous involved heat treatment procedure.

Table 3. Mechanical Properties of Haynes™ HR 120™ alloy						
Condition	YS MPa	UTS MPa	E%	HBW	SSR@889	SSR@982
S1	343	745	36	202	120 hr*	110 hr**
S2	336	750	41	205	120 hr*	110 hr**
S3	333	740	45	194	120 hr*	110 hr**

*Tests are interrupted after a life time of 120 hr.
**Tests are interrupted after a life time of 110 hr.

Microstructural characteristics

The effects of solution heat treatments on the microstructural characteristics of alloy Haynes™ HR-120™ are reported in Table 4. The first parameter to evaluate is the microstructural uniformity in transversal sections. The evaluation of as-forged samples show non-uniform structures per ASTM E1181, a duplex wide range structure is found with the following characteristics: 70% with an average grain size of 75.5µm ± 10 (ASTM 4.5) and a grain size ALA 359.2 ± 30 µm (ASTM 0). Fig. 1 shows this microstructure.

When this initial microstructure is exposed to the solution treatment S1, no significant changes are observed. Some recrystallization is found since 80% of the microstructure consists of grains with an average size of 37.8 µm ± 6 (ASTM 6.5) and remaining ALA average grain size of 359.2 ± 30 µm (ASTM 0). A representative view of the microstructure is shown as Figure 2. This structure after indicates that heat treatment at 1000°C is not sufficient to promote a full re-crystallization of the initial microstructure. Most likely a longer soaking period would be needed to promote a complete recrystallization.

Table 4. Microstructural characteristics of Haynes™ HR 120™ alloy

Condition	Uniformity per ASTM E1181	Avg. Grain size μm	ALA Grain Size μm
As Forged	Duplex, wide range	75.5 ± 10	359.2 ± 30
S1	Duplex, wide range	37.8 ± 6	359.2 ± 30
S2	Uniform	31.8 ± 5	127.0 ± 20
S3	Uniform	63.5 ± 10	254.0 ± 40

Recrystallization is promoted in alloy Haynes™ HR-120™ during heat treatment S2. The resulting microstructure consists of an average grain size of 31.8μm ± 6 (ASTM 7.0). The original ALA grains are also re-crystallized to generate new ALA grain with an average size of 127 ± 20 μm (ASTM 3.0). These solution temperature and soaking parameters are sufficient to promote a uniform microstructure as desired (see Figure 3 for a representative view). A similar behavior is observed for samples exposed to treatment S3 which includes a temperature above 1100°C. In this case, a uniform and fully recrystallized microstructure is promoted, but some grain coarsening is also found. A representative microstructure exposed to this heat treatment is shown as Figure 4 with the grain size characteristics in Table 4.

Considering this information and results, the most appropriate microstructure is promoted when alloy Haynes™ HR-120™ is exposed to the solution heat treating conditions S2.

Figure 1: Representative microstructure of alloy Haynes™ HR-120™ in the as forged condition. A duplex, wide range structure is found.

Figure Figure 2: Microstructure of alloy Haynes™ HR-120™ exposed to Solution S1 conditions. A duplex, wide range structure is found together with partial recrystallization.

Microstructural evaluation by SEM

In order to evaluate in detail the effect of microstructure on the described mechanical and microstructural properties, SEM evaluations are performed in non-stressed ends of tensile specimens. Results are summarized as follows: for the condition S1, the resulting microstructure is duplex and wide range, indicating a limited re-crystallization and the presence of intergranular precipitates. Such precipitates, containing by Ni-Fe-Cr-Nb and Ni-Fe-Cr are promoted during

166

melting operations when the alloy is produced. A representative image of this microstructure is shown as Figure 5.

Figure 3: Representative microstructure of alloy Haynes™ HR-120™ exposed to treatment S2 conditions. A uniform fully re-crystallized microstructure is observed.

Figure 4: Representative microstructure of alloy Haynes™ HR-120™ exposed to Solution S3 conditions.

The number of intergranular precipitates is reduced for the heat treatment S2. As consequence of this, the elongation properties are improved. As mentioned before, another important effect of these solution parameters is the promotion of recrystallization. Figure 6 presents a representative image of the microstructure after treatment S2. Finally, when alloy Haynes™ HR-120™ is exposed to Solution S3 the intergranular precipitates are dissolved. One of the main consequences is that grain coarsening takes place more rapidly since intergranular precipitates act as a limiting barrier for this phenomenon. On the other hand, the elongation properties show the best results for this condition. Figure 7 shows a representative image of the promoted microstructure.

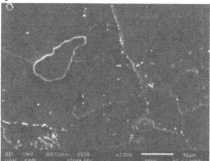

Figure 5: Representative tensile specimen after treatment S1. Inter-granular precipitates are present.

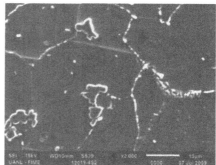

Figure 6: Representative tensile after treatment S2. Presence of intergranular precipitates is reduced

increased, the presence of intergranular precipitates is reduced. These microstructural observations allow explanation for the elongation properties here reported since the presence of such precipitates can promote a reduction of this tensile characteristic.

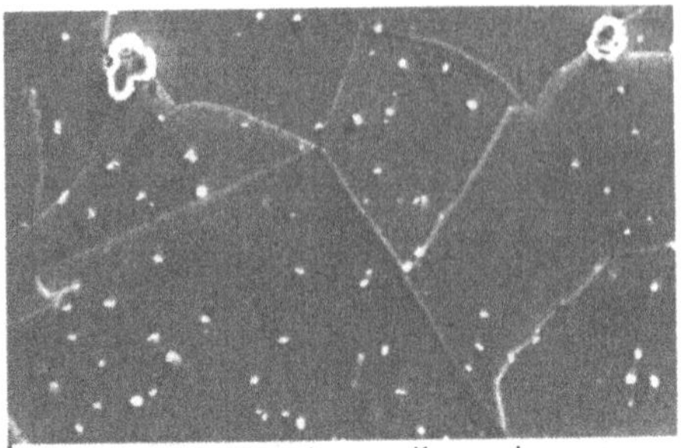

CONCLUSIONS

Alloy Haynes^{TM} HR-120^{TM} has been exposed to different solution temperatures around 1100°C, promoting differences of mechanical properties. The presence of intergranular precipitates, formed during melting operations, can be dissolved for a solution temperature above 1050°C. Dissolution of intergranular precipitates promotes better elongation properties on for tensile testing. Such dissolution of intergranular precipitates allows grain coarsening above 1100°C. Effective recrystallization can be promoted for solution temperatures above 1050°C

Figure 7: Representative tensile specimen exposed to treatment S3. Grain boundaries are free of intergranular precipitates.

REFERENCES

1. Haynes HR-120® alloy brochure, Haynes international, 1992, USA.
2. Superalloys, M.J. Donachie and S.J. Donachie, ASM International, 2003, USA.
3. Gas Turbine Handbook, A. Giampaolo, The Fairmont Press Inc., 2006, USA.
4. Wrougth And Cast Heat Resistant Stainless Steels An Nickel Alloys For The Refining And Petrochemical Industries, D.J. Tillack and J.E. Guthrie, Nickel Institute, 2007, USA.
5. Fabrication of Haynes and Hastelloy® Solid-Solution-Strengthened High-Temperature Alloys, Haynes International, 2002, USA.

AUTHOR INDEX

SUBJECT INDEX

Printed in the United States
By Bookmasters